EDEXCEL
A LEVEL MATHS
YEAR 1 + YEAR 2

MECHANICS STUDENT WORKBOOK

Author
Steve Cavill

Powered by **MyMaths**.co.uk

OXFORD
UNIVERSITY PRESS

Great Clarendon Street, Oxford, OX2 6DP, United Kingdom

Oxford University Press is a department of the University of Oxford.

It furthers the University's objective of excellence in research, scholarship, and education by publishing worldwide. Oxford is a registered trade mark of Oxford University Press in the UK and in certain other countries

British Library Cataloguing in Publication Data

Data available

978-0-19-841326-4

3 5 7 9 10 8 6 4

Paper used in the production of this book is a natural, recyclable product made from wood grown in sustainable forests.

The manufacturing process conforms to the environmental regulations of the country of origin.

Printed in Great Britain by CPI Group (UK) Ltd., Croydon CR0 4YY

Acknowledgements

Author
Steve Cavill

Editorial team
Dom Holdsworth, Ian Knowles, Matteo Orsini Jones

With thanks to
Katherine Bird, Amy Ekins-Coward, Neil Tully

Although we have made every effort to trace and contact all copyright holders before publication this has not been possible in all cases. If notified, the publisher will rectify any errors or omissions at the earliest opportunity.

Links to third party websites are provided by Oxford in good faith and for information only. Oxford disclaims any responsibility for the materials contained in any third party website referenced in this work.

Contents

About this book

This book is designed to complement the Student Books in the series *Edexcel A Level Maths* and provides you with extra support for mechanics. An introductory chapter offers advice on modelling real-world situations using mechanics; indicates how the various topics within mechanics are related to one another; and provides a summary of the relevant notation and language required. There is then one section in this book for every mechanics section in the Student Book. Each of these sections provide a recap of the main ideas in the Student Book and illustrates their application with examples of exam-style questions and model student answers, followed by three pages of exam-style questions for you to attempt yourself.

Full details of the mechanics content of the AS level (8MA0) and A level (9MA0) exam specifications can be found on the Edexcel web site.

https://qualifications.pearson.com/en/qualifications/edexcel-a-levels/mathematics-2017.html

At AS Level, mechanics is tested in section B of paper 2; there are 30 marks and approximately 37.5 minutes available. At A Level, mechanics is tested in section B of paper 3; there are 50 marks and approximately 60 minutes available.

Answers

The back of this book contains short answers to all the questions. Full mark schemes for each question can be found online, with password protection for teachers.

https://global.oup.com/education/content/secondary/series/edexcelalevelmaths-answers

Formulae

In the exam, you will be provided with a 'Mathematical Formulae and Statistical Tables' booklet. The relevant mechanics formulae are provided at the end of this book.

Calculators

All papers are calculator papers. You must make sure that you have a calculator and that you know how to use it. The rules on which calculators are allowed can be found in the Joint Council for General Qualifications document 'Instructions for conducting examinations' (ICE).

Introduction to mechanics

Mathematical modelling

This introduction to mechanics puts a lot of focus on the notion of **mathematical modelling**. You might have only met the word *model* as something that's a miniature of a life-size object, such as a toy car or a doll's house, or perhaps you've come across models in your studies of Design and Technology. As such, you might be new to the idea of a model being an equation, or set of equations, that aims to predict a real-life scenario is likely to be new to them.

What is Mechanics?

Like statistics, mechanics is a branch of applied Mathematics. Mechanics helps us to understand and predict the behaviour of objects that are subjected to forces. Mechanics is subdivided into *dynamics*, which studies moving situations, and *statics*, which studies stationary situations. Note, be careful with the spelling of *statics* and *statistics* – they may look similar, but they have very different meanings.

Think about whether these situations would classify as dynamics or statics.

- The motion of an aircraft.
- The working of a car engine.
- The stresses and strains in a bridge.
- The stability of a building.
- The movement of a lift inside a building.
- The structure of a television mast.

These are all examples of situations that mechanics attempts to analyse. But these and other situations are complex and must first be simplified by the process of mathematical modelling.

What is mathematical modelling?

Put simply, Mechanics aims to simplify real-life situations by creating mathematical models, which are then used to predict outcomes. The model's outcomes can then be compared with those of the real-life situation. The model takes the form of an equation or a set of equations that link inputs to outcomes. This process gives a measure of the success of the model. If the comparison of the real-life outcomes with the model's outcomes is good, then the model is a good one; if not, then the model needs to be modified.

This flow-chart summarises the modelling process.

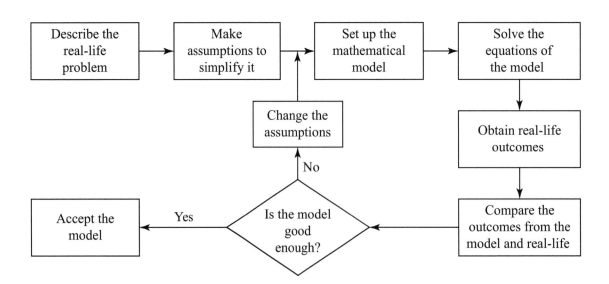

Note that simplifying the real-life situation requires making assumptions about which factors might affect the outcomes and which of these factors can be ignored.

For example, imagine a lift carrying people up a tall building. What factors do you think might affect the working of the lift? Which of these factors are more important than others, and which might be simplified or neglected by a model?

Once you have decided all these factors, the mathematical model then provides a set of equations that connects the relevant factors and ignores the others.

Try these questions

1   Do you think it would be reasonable to ignore resistance to motion in the following situations?

   a   A rock is dropped from an upstairs window.

   b   A hockey puck slides across an ice rink.

   c   An inflated beach ball is dropped from an upstairs window.

   d   A bucket is lifted from a well over a smooth pulley.

   e   A motorboat drives through water.

   f   A person walks at $1 \, \text{m s}^{-1}$

   g   A person runs at $8 \, \text{m s}^{-1}$

   h   A person coasts down a steep hill on a bicycle.

2   You are given a question with the following wording.
   "A snooker ball, moving at a constant speed on a smooth table, strikes another ball at rest. The first ball comes to a stop and the second ball moves away with a speed equal to the first ball's initial speed."

   a   What assumptions are being made in this question?

   b   How do you think the real-life situation would differ?

3   You are given a question with the following wording.
   "A rocket, of mass 2000 kg, is launched from Earth. During the launch, the engine provides a constant upwards thrust of 50 000 N. How fast will the rocket be moving after 10 seconds?

   a   What assumptions are being made in this question?

   b   For each assumption, describe how the situation would differ in real life and explain how this might affect the final answer.

4   You are given a question about a system involving three masses, all connected by strings over pulleys, with one mass resting on a smooth table and the others suspended freely either side of the table.

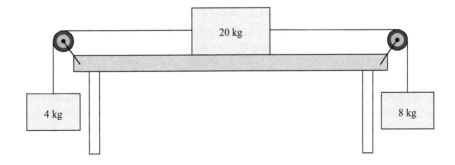

   a   What assumptions would you say are reasonable to make in this model?

   b   Are there any assumptions that you don't think are valid for this model?

   c   If the assumptions described in part **a** were not true, how would the real-life situation differ?

Language and notation in mechanics

In mechanics, certain words are used to indicate that a situation has been simplified. These words are used in problems to help you formulate the situation and define the mathematical model.

Noun	Applied to	Simplification/meaning
Particle	Objects	Size is negligible
Rod	Beams, struts, joists	Width is negligible and rod is rigid
String	Wires, cables	Width and weight are negligible
Lamina	Flat sheets of material	Thickness is negligible; sheet is rigid

Adjective	Applied to	Simplification/meaning
Light	Strings, springs, rods, pulleys	Mass is negligible
Small	Objects	Size is negligible
Inextensible	Strings, rods	No stretching occurs
Rigid	Objects, rods, structures	No bending or deformation occurs
Smooth	Surfaces	Resistance to motion is negligible
Rough	Surfaces	Resistance to motion must be taken into account
Taut	Strings	Not slack; tightened
Thin	Laminas	Thickness is negligible
Instantaneous	Events	Events take no time to occur
Uniform	Objects, motion	Equivalent to constant

The following units and symbols are taken from Ofqual guidance.

kg	kilogram	m	metre
km	kilometre	m/s, $m\,s^{-1}$	metres per second (velocity)
m/s^2, $m\,s^{-2}$	metres per second per second (acceleration)		
F	Force or resultant force	N	newton
N m	newton metre (moment of a force)	t	time
s	displacement	u	initial speed
v	final speed	a	acceleration
g	acceleration due to gravity	μ	coefficient of friction

Diagrams in mechanics

Almost every problem posed in mechanics is expressed in words. Almost every solution to a problem is helped by a good-quality diagram. A diagram that accompanies a written solution can provide a significant part of the solution. It serves several purposes: it can

- Provide visual support to help you imagine the scenario of the problem,
- Display all the information, usually numerical values, that are contained in the text of the problem. Drawing a diagram means you no longer have to refer back to the question repeatedly, in search of numerical values.
- Provide labels, such as for points, angles and forces, that you can refer to in your written solution.

A diagram should not be cramped, as it needs to be large enough to display all the required information in the question.

Forces, displacements, velocities and accelerations are all vectors that can appear on diagrams as labelled arrows. It is usual that the arrows for different vectors have different styles of arrowheads: for example, arrows representing forces are different to those for velocities and accelerations.

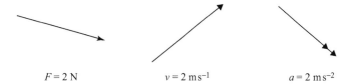

$$F = 2\,\text{N} \qquad v = 2\,\text{m}\,\text{s}^{-1} \qquad a = 2\,\text{m}\,\text{s}^{-2}$$

Whilst it can be difficult to draw different arrowheads by hand, drawing a double arrowhead for acceleration is good practice –as long as you use whatever system you choose consistently.

Vectors representing forces, displacements, velocities and accelerations are usually printed in books in bold, such as **F**. When you write them by hand, you should underline the vector with a straight line or, preferably, a 'twiddle' (officially called a tilde), such as F̰

Where a problem involves, say, two objects in contact with each other, the vectors associated with the two objects may not be clearly allocated to them if the objects are sketched in contact. For example, in diagram **a**, it is not clear which of the two forces are acting on which of the colliding spheres P and Q. However, when the spheres are drawn slightly separated as in diagram **b**, there is no confusion. Such a diagram is called an *exploded diagram*.

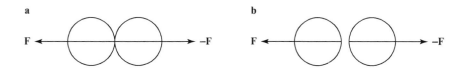

Inter-relationships between topics

There are many inter-relationships within mechanics, and between mechanics and other topics, that give opportunity for synoptic assessment.

The diagram shows a number of connections between different topics covered in this course, though this is by no means an exhaustive list and you should always look for connections to topics you've already learnt.

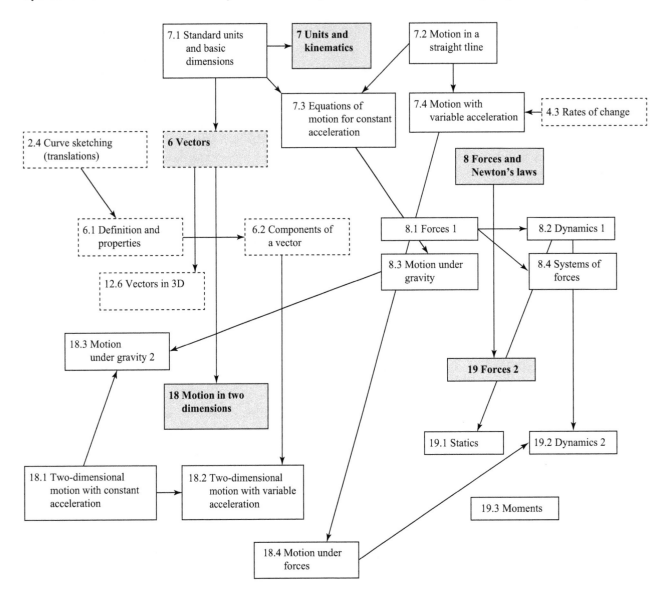

Recap

- SI units (*Système international d'unités*) are an international system of units of measurement which are built on seven **base units** which can be combined to give **derived units**.
- In **mechanics** three base units are used
  - kilograms (kg) for measuring mass
  - metres (m) for measuring length or distance
  - seconds (s) for measuring time
- The most common derived units in mechanics are
  - metres per second ($m\,s^{-1}$) for measuring speed
  - metres per second squared ($m\,s^{-2}$) for measuring acceleration
  - newtons (N or $kg\,m\,s^{-2}$) for measuring force

> Newton's second law tells us that Force = mass × acceleration so the unit of force, the newton, comes from mass (kg) × acceleration ($m\,s^{-2}$)

> If you think it is strange talking about a 'square second' or 'seconds squared' think about acceleration as *'metres per second per second'*—how much the velocity changes each second.

- Some quantities use different words if they are describing a **vector** or **scalar**.

Scalar quantities		Vector quantities	
distance	for example, 20 m	displacement	for example, 20 m left
speed	for example, $72\,m\,s^{-1}$	velocity	for example, $72\,m\,s^{-1}$ north-east
acceleration	for example, $9.8\,m\,s^{-2}$	acceleration	for example, $9.8\,m\,s^{-2}$ downwards

- There are standard prefixes which can be applied to all SI units.

'kilo' means one thousand	A **kilo**gram is 1000 grams
	A **kilo**metre is 1000 metres
'centi' means one hundredth	There are 100 **centi**metres in a metre
'milli' means one thousandth	There are 1000 **milli**metres in a metre

You are probably familiar with using these prefixes with grams and metres, but they can be applied to any SI unit. For example, there are 1000 mN in 1 N.

- You will need to be able to convert between different units.
  The most common conversion is to change speeds in $m\,s^{-1}$ into $km\,h^{-1}$ and vice-versa.

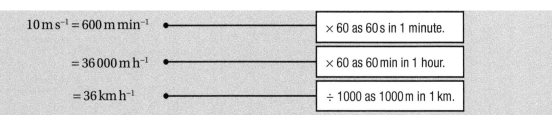

$$10\,m\,s^{-1} = 600\,m\,min^{-1}$$ ← × 60 as 60 s in 1 minute.

$$= 36\,000\,m\,h^{-1}$$ ← × 60 as 60 min in 1 hour.

$$= 36\,km\,h^{-1}$$ ← ÷ 1000 as 1000 m in 1 km.

Many people simply remember the conversion factor of 3.6 but you may still need the steps above (or similar) if the question deals with less common units.

Example 1

An adult snail can move at a speed of 1 millimetre per second.

Find its speed in $km\,h^{-1}$

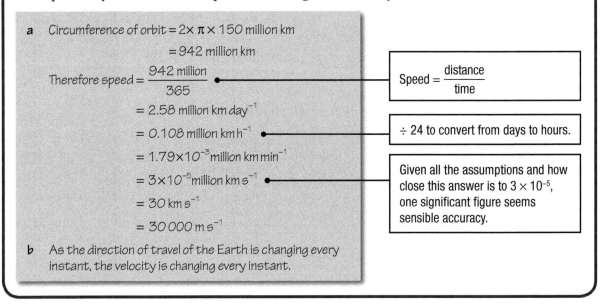

$1\,mm\,s^{-1}=0.001\,m\,s^{-1}$ ———————————————— There are 1000 mm in 1 m

$\quad\quad\quad = 0.0036\,km\,h^{-1}$ ———————————————— $\times\,3.6$ to convert $m\,s^{-1}$ into $km\,h^{-1}$

Example 2

Assume that the Earth

- takes 365 days to travel round the Sun
- has a circular orbit of radius 150 million km
- moves at constant speed.

a Calculate the speed of the Earth in $m\,s^{-1}$

b Explain why the calculation in part **a** does not give the *velocity* of the Earth.

a Circumference of orbit $= 2\times\pi\times 150$ million km

$\quad\quad\quad\quad\quad\quad\quad = 942$ million km

Therefore speed $= \dfrac{942\,\text{million}}{365}$ ———————— $\text{Speed} = \dfrac{\text{distance}}{\text{time}}$

$\quad\quad\quad = 2.58$ million $km\,day^{-1}$

$\quad\quad\quad = 0.108$ million $km\,h^{-1}$ ———————— $\div\,24$ to convert from days to hours.

$\quad\quad\quad = 1.79\times 10^{-3}$ million $km\,min^{-1}$

$\quad\quad\quad = 3\times 10^{-5}$ million $km\,s^{-1}$ ———————— Given all the assumptions and how close this answer is to 3×10^{-5}, one significant figure seems sensible accuracy.

$\quad\quad\quad = 30\,km\,s^{-1}$

$\quad\quad\quad = 30\,000\,m\,s^{-1}$

b As the direction of travel of the Earth is changing every instant, the velocity is changing every instant.

Exam tips

- Think carefully whether to multiply or divide when converting units and check whether your answer is sensible. For example, when changing from $km\,h^{-1}$ to $km\,s^{-1}$, ask yourself, 'Will the number of km travelled be bigger or smaller?'
- Check whether you are working with scalar or vector quantities and if your answer is a vector remember to give the direction.
- Make sure all your units are consistent before carrying out calculations.

Exam practice questions

1 Which of these units is it most appropriate to use when describing the weight of an adult woman?

kg g kN N

Explain your answer. [1]

2 Change $530\,\text{km}\,\text{h}^{-1}$ into $\text{m}\,\text{s}^{-1}$ [2]

3 There are 1000 micrometres (μm) in 1 millimetre.

Convert $17\,000\,\mu\text{m}\,\text{s}^{-1}$ into $\text{km}\,\text{h}^{-1}$ [3]

4 The Lyke Wake Walk is a 40 mile walk across the North Yorkshire Moors.

Steve started the walk at 11 pm and completed it at 4.30 pm the following day.

Calculate his average walking speed in metres per second. You can assume that 1 mile = 1.6 km [3]

5 A train accelerates uniformly from $u = 20\,\text{km}\,\text{h}^{-1}$ to $v = 80\,\text{km}\,\text{h}^{-1}$ over 4 minutes.

    a Calculate the acceleration of the train in $\text{m}\,\text{s}^{-2}$

       (You may use the formula $v = u + at$)   [3]

    b How many kilometres does the train travel while accelerating?

       (You may use the formula $v^2 = u^2 + 2as$)   [3]

Recap

These terms are used to describe how things move.

- **Position** is a vector – it gives distance and direction from an origin.
- **Displacement** is also a vector – it describes a change in position.
- **Distance** is the scalar equivalent of displacement.

The position and displacement at point A is $-3\,\text{m}$, and the distance from O is $3\,\text{m}$.

If a particle moves from O to A then to B, its final displacement from O is $7\,\text{m}$ but the distance it has travelled is $13\,\text{m}$.

- **Velocity** is a vector – it is the rate of change of displacement.
- **Speed** is the scalar equivalent of velocity.

$$\text{Average speed} = \frac{\text{Total distance}}{\text{Total time}} \qquad \text{Average velocity} = \frac{\text{Resultant displacement}}{\text{Total time}}$$

If the motion described above took 10 seconds

$$\text{Average speed} = \frac{\text{Total distance}}{\text{Total time}} \qquad \text{Average velocity} = \frac{\text{Resultant displacement}}{\text{Total time}}$$

$$= \frac{13}{10} \qquad\qquad\qquad = \frac{7}{10}$$

$$= 1.3\,\text{m s}^{-1} \qquad\qquad\quad = 0.7\,\text{m s}^{-1}$$

- You can illustrate motion using kinematic graphs. Time (t) is always plotted on the horizontal axis.
- On a displacement–time (s–t) graph
  - **gradient** represents **velocity**.

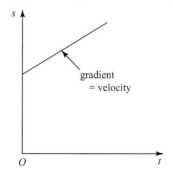

- On a velocity–time (v–t) graph
  - **gradient** represents **acceleration**
  - **area** (between graph and t-axis) represents **displacement**.

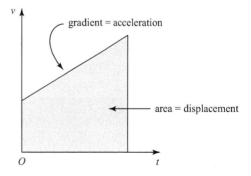

Example 1

This velocity–time graph represents the motion of a particle for 80 seconds.

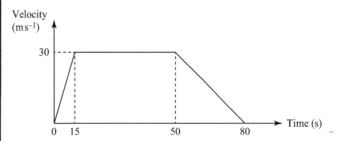

a Calculate the initial acceleration of the particle.

b Calculate the distance that the particle travels over the 80 seconds.

c Find the average velocity of the particle over the 80 seconds.

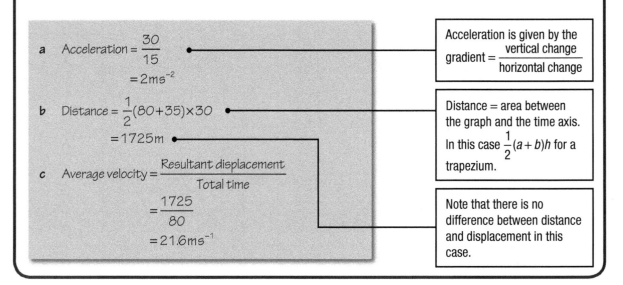

a Acceleration $= \dfrac{30}{15}$

$= 2\,\text{ms}^{-2}$

b Distance $= \dfrac{1}{2}(80+35)\times 30$

$= 1725\,\text{m}$

c Average velocity $= \dfrac{\text{Resultant displacement}}{\text{Total time}}$

$= \dfrac{1725}{80}$

$= 21.6\,\text{ms}^{-1}$

Acceleration is given by the gradient $= \dfrac{\text{vertical change}}{\text{horizontal change}}$

Distance = area between the graph and the time axis.
In this case $\dfrac{1}{2}(a+b)h$ for a trapezium.

Note that there is no difference between distance and displacement in this case.

Example 2

A girl throws a pebble vertically upwards. Given that the acceleration of the pebble is constant and acts down, sketch a velocity-time graph to show the motion of the pebble until it hits the ground.

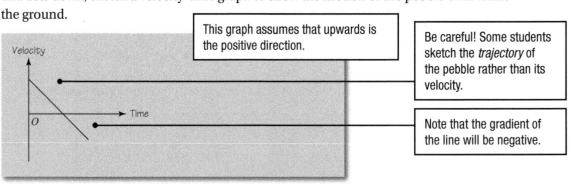

This graph assumes that upwards is the positive direction.

Be careful! Some students sketch the *trajectory* of the pebble rather than its velocity.

Note that the gradient of the line will be negative.

Exam tips

- Check the labels on the axes so you know if you are dealing with a v–t graph or an s–t graph, or occasionally even an acceleration–time graph.
- If you are calculating vector quantities make sure you include the direction.
  Remember acceleration may be a scalar or a vector.
- Draw graphs large enough to be clear and be ready to redraw them if you need to after calculating some values. Make sure you label all the velocities or distances and times on the axes.

1   This displacement–time graph represents the motion of a car.

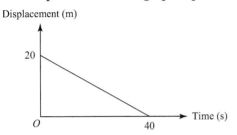

Displacement (m)

20

O

40

Time (s)

Calculate the velocity of the car.   [2]

2   A particle has initial velocity of $16\,\mathrm{m\,s^{-1}}$ and it moves with

- deceleration $2\,\mathrm{m\,s^{-2}}$ for 5 seconds
- constant velocity for the next 5 seconds
- deceleration $1\,\mathrm{m\,s^{-2}}$ for the next 20 seconds

a   Sketch a velocity–time graph.

Mark on the axes any values of v and t where the acceleration changes.   [4]

b   Calculate the particle's displacement from its starting position after it has finished the motion.   [4]

c   Calculate the average velocity of the particle.   [2]

3   A car is travelling at a constant velocity of $20\,\text{m s}^{-1}$ when it passes a stationary motorcycle.

5 seconds later the motorcycle starts accelerating at $2\,\text{m s}^{-2}$ until it catches up with the car T seconds after the car first passed it.

a   On the same axes, sketch velocity–time graphs to represent the motions of the car and the motorcycle for $0 \leq t \leq T$ [3]

b   Show that $T^2 - 30T + 25 = 0$ [4]

c   Find how far the motorcycle has travelled when it catches the car.   [3]

Recap

- The variables used in equations for motion under constant acceleration are
  - s – displacement
  - v – final velocity
  - t – time
  - u – initial velocity
  - a – acceleration

- There are five equations for motion with constant acceleration.
  Each one uses four out of the five variables.
  They are often called the *suvat* or *uvast* equations.

  - $v = u + at$
  - $s = \dfrac{1}{2}(u+v)t$
  - $s = ut + \dfrac{1}{2}at^2$

  - $v^2 = u^2 + 2as$
  - $s = vt - \dfrac{1}{2}at^2$

- These equations are derived using the features of kinematic graphs. This graph represents a particle accelerating from initial velocity u to final velocity v over a time t (mostly units will be SI but any consistent units could be used).

- The straight line shows that the acceleration is constant and the value of that acceleration is the gradient of the graph.

- Writing an equation for the gradient and rearranging it produces the first *suvat* equation.

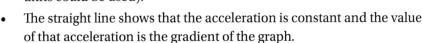

$$\text{gradient} = \frac{\text{vertical change}}{\text{horizontal change}} \Rightarrow a = \frac{v-u}{t}$$

$$at = v - u$$

$$v = u + at$$

- The area between the graph and the t-axis is equal to the displacement. This gives the second *suvat* equation.

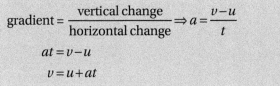

$$\text{Area} = \frac{1}{2}(a+b)h \quad \Rightarrow \quad s = \frac{1}{2}(u+v)t$$

- The other three equations are found by substituting from one of these into the other to eliminate v, or u or t.

For example, $v = u + at \Rightarrow t = \dfrac{v-u}{a}$

$$s = \frac{1}{2}(u+v)t$$

$$s = \frac{1}{2}(u+v)\frac{(v-u)}{a}$$

$$2as = (v+u)(v-u) = v^2 - u^2$$

$$v^2 = u^2 + 2as$$

Example 1

A particle accelerates uniformly at $4\,\text{m s}^{-2}$ to a final velocity of $10\,\text{m s}^{-1}$ while travelling $8\,\text{m}$

What was its starting velocity?

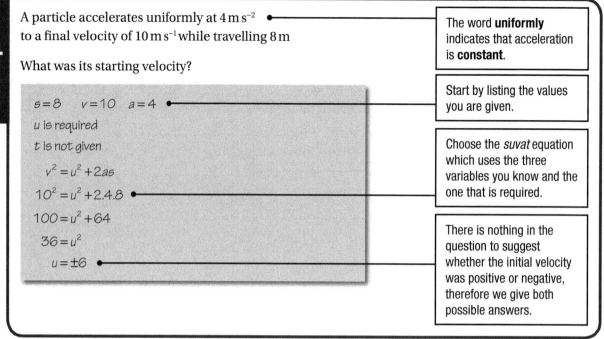

$s=8 \quad v=10 \quad a=4$

u is required

t is not given

$v^2 = u^2 + 2as$

$10^2 = u^2 + 2.4.8$

$100 = u^2 + 64$

$36 = u^2$

$u = \pm 6$

The word **uniformly** indicates that acceleration is **constant**.

Start by listing the values you are given.

Choose the *suvat* equation which uses the three variables you know and the one that is required.

There is nothing in the question to suggest whether the initial velocity was positive or negative, therefore we give both possible answers.

Example 2

Particle P starts from the origin moving with initial velocity $6\,\text{m s}^{-1}$ and acceleration $2\,\text{m s}^{-2}$

a Write an expression, in terms of t, for the displacement of particle P after t seconds.

Particle Q starts from displacement $60\,\text{m}$ with initial velocity $9\,\text{m s}^{-1}$ and acceleration $-4\,\text{m s}^{-2}$

b After how long will the particles be at the same place?

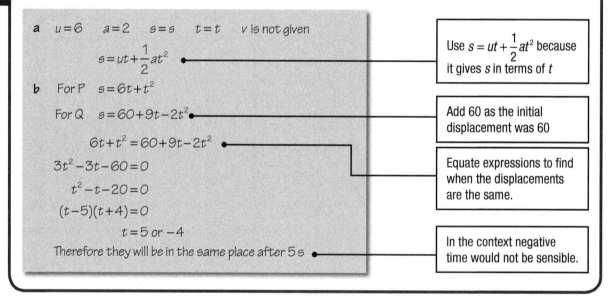

a $\quad u=6 \quad a=2 \quad s=s \quad t=t \quad v$ is not given

$$s = ut + \frac{1}{2}at^2$$

b For P $\quad s = 6t + t^2$

For Q $\quad s = 60 + 9t - 2t^2$

$6t + t^2 = 60 + 9t - 2t^2$

$3t^2 - 3t - 60 = 0$

$t^2 - t - 20 = 0$

$(t-5)(t+4) = 0$

$t = 5$ or -4

Therefore they will be in the same place after $5\,\text{s}$

Use $s = ut + \dfrac{1}{2}at^2$ because it gives s in terms of t

Add 60 as the initial displacement was 60

Equate expressions to find when the displacements are the same.

In the context negative time would not be sensible.

Exam tips

- Make sure you have constant acceleration before trying to use *suvat* equations.
- Write down the values you are given for s, u, v, a and t and also list what you are trying to find to help you choose the right *suvat* equation to use.
- If there is a choice of *suvat* equations, choose the one where you use given values and not calculated values.
- Think about what a negative answer means for a displacement or velocity and remember to give your answers in the context of the question where necessary.

1   $u = 10$, $a = 5$ and $t = 12$. Find v [2]

2   A particle starts with velocity $12\,\mathrm{m\,s^{-1}}$ and accelerates at $3\,\mathrm{m\,s^{-2}}$ for one minute.

   a   What is its final velocity? [3]

   b   How far has the particle travelled during this minute? [4]

3   A car decelerates at $2\,\mathrm{m\,s^{-2}}$ and comes to rest in $100\,\mathrm{m}$.

   How fast was the car travelling before the deceleration? [3]

4 A particle has initial velocity of $u \, \text{m s}^{-1}$ and accelerates at $a \, \text{m s}^{-2}$

a Given that the particle travels 324 m in 8 s, find an equation relating a and u. [2]

b In the next 52 seconds it travels a further 13 km and 26 m. Find u and a [6]

5 Use this diagram to help explain why $v = u + at$ [3]

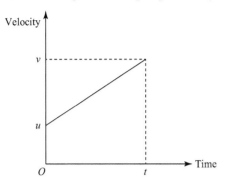

6 A car takes 24 s to travel 648 m during which time its velocity doubles.

Find its initial velocity. [4]

Recap

- You know from Section 7.3 that acceleration is equal to the **gradient** of a straight-line velocity–time graph. This is also true for a curved velocity–time graph.
- You also know that the gradient of a displacement–time graph is equal to velocity.
- You can find the gradient by differentiating.

  - $a = \dfrac{dv}{dt}$

  - $v = \dfrac{ds}{dt}$

- You saw that displacement is equal to the **area** under a velocity–time graph.
- You can find the area under a curved graph by integrating, which is also the reverse of differentiating.

  - $s = \int v \, dt$

  - $v = \int a \, dt$

- In kinematics, you will often see the notation $\dot{s}$ for $\dfrac{ds}{dt}$ and $\ddot{s}$ for $\dfrac{d^2 s}{dt^2}$

  - $v = \dot{s} = \dfrac{ds}{dt}$

  - $a = \dot{v} = \dfrac{dv}{dt} = \ddot{s} = \dfrac{d^2 s}{dt^2}$

- To find the displacement over a period of time you use definite integration.

For example, to find the displacement from $t = 3$ to $t = 7$ you calculate $\int_3^7 v(t) \, dt$

Example 1

The displacement of a particle from the origin is given by $s = t^3 + 4t^2 - 5t$

Find

The formula for s in terms of t is a good indicator that you will need to use calculus.

a   An expression for the velocity, v at time t,

b   The acceleration after 5 seconds.

a $s = t^3 + 4t^2 - 5t$

     $v = 3t^2 + 8t - 5$ Differentiate s to find v

b $a = 6t + 8$ Differentiate v to find a

     $= 6 \times 5 + 8$ Substitute $t = 5$

     $= 38 \, ms^{-2}$

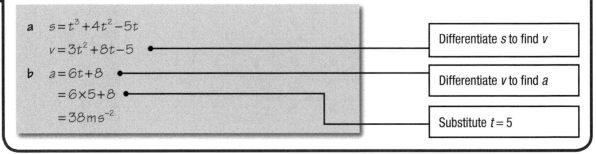

Example 2

Particle P starts from the origin with velocity $10\,\text{m s}^{-1}$ and constant acceleration $-3\,\text{m s}^{-2}$

Particle Q also starts from the origin with velocity $2\,\text{m s}^{-1}$ and its acceleration at time t is given by $a = 2t - 5$

After how long will the particles be in the same place again?

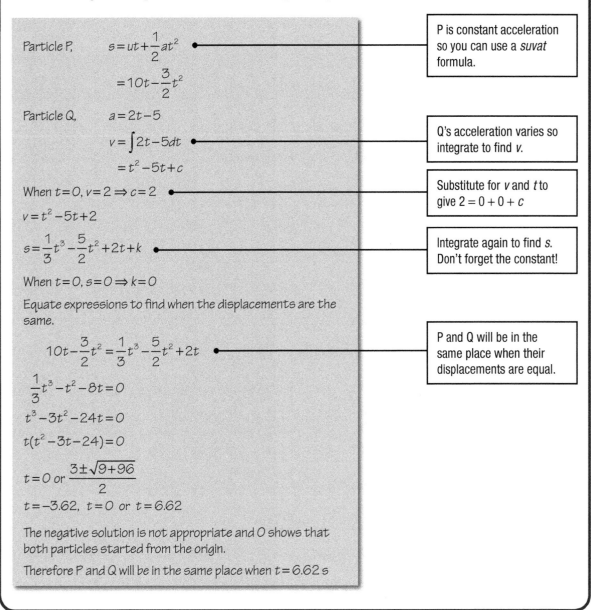

Particle P, $s = ut + \dfrac{1}{2}at^2$

$= 10t - \dfrac{3}{2}t^2$

> P is constant acceleration so you can use a *suvat* formula.

Particle Q, $a = 2t - 5$

$v = \int 2t - 5\,dt$

$= t^2 - 5t + c$

> Q's acceleration varies so integrate to find v.

When $t = 0$, $v = 2 \Rightarrow c = 2$

> Substitute for v and t to give $2 = 0 + 0 + c$

$v = t^2 - 5t + 2$

$s = \dfrac{1}{3}t^3 - \dfrac{5}{2}t^2 + 2t + k$

> Integrate again to find s. Don't forget the constant!

When $t = 0$, $s = 0 \Rightarrow k = 0$

Equate expressions to find when the displacements are the same.

$10t - \dfrac{3}{2}t^2 = \dfrac{1}{3}t^3 - \dfrac{5}{2}t^2 + 2t$

> P and Q will be in the same place when their displacements are equal.

$\dfrac{1}{3}t^3 - t^2 - 8t = 0$

$t^3 - 3t^2 - 24t = 0$

$t(t^2 - 3t - 24) = 0$

$t = 0$ or $\dfrac{3 \pm \sqrt{9 + 96}}{2}$

$t = -3.62$, $t = 0$ or $t = 6.62$

The negative solution is not appropriate and 0 shows that both particles started from the origin.

Therefore P and Q will be in the same place when $t = 6.62\,\text{s}$

Exam tips

- If you are given expressions for s, v or a in terms of t this should alert you to use calculus and not the *suvat* equations.
- Don't forget the constant when integrating, it may be 0, but you should show that you have remembered it.
- Remember that if velocity becomes negative, then to calculate distance travelled rather than displacement you will need to split the integral.
  Sketching a graph of the motion will help you see what is happening.

1   The displacement in metres of a particle at time t seconds is $5t^4 - 3t^2 + 4$

What is its velocity in m s^{-1} after t seconds?   [2]

2   The displacement, s metres, of a particle at time t is given by $s = 6 + 5t + 12t^2 - t^3$

    a   Find an expression for the velocity of the particle at time t [2]

    b   At what value of t is the acceleration of the particle 0?   [2]

3   The displacement of a particle is given by $s = t^3 - 18t^2 + 81t + 20$

How many seconds pass between the times when the particle is stationary?   [4]

4 The velocity of a particle is $v = 3t^2 - \dfrac{3}{t^2}$

    a How far does the particle travel from $t = 1$ to $t = 5$?   [4]

    b Why is this model not appropriate for all values of t?   [1]

5 The velocity of a particle at time t s is given by $v = 6t^{\frac{1}{2}}$ m s^{-1}

The displacement is 100 m when the acceleration is 1 m s^{-2} and the acceleration and displacement are numerically equal at time T

Show that $T^2 = \dfrac{3 + 8\sqrt{T}}{4}$ [7]

6 Given that $\ddot{s} = a$ and that $\dot{s} = u$ when $t = 0$, use integration to prove that $s = ut + \dfrac{1}{2}at^2$

What assumption is made for this proof to work? [4]

Assumption

Recap

A **force** is a **push** or **pull**.
Its effect is to either start something moving or alter the motion of an already moving body.

- In Mechanics, you meet several different kinds of forces.
  - A rod under **tension** can exert a pull or under **thrust** can exert a push.
  - A string under tension can exert a pull.
    (A string cannot exert a push.)
  - Gravity acts on any massive body to give a vertically downwards force – **weight**; see section **8.3**
  - Two bodies in contact that press against one another give rise to a **normal reaction**.
    The direction of this force is at right angles, that is, normal to the surface.
  - **Friction** is the force exerted by a 'rough surface' that acts to stop a body moving.
    If a surface exerts no or negligible friction you say that it is 'smooth'.
  - An engine, or motor, can exert a driving force.

Newton's first law

A body will continue in a state of **equilibrium** unless acted upon by a force.

Equilibrium means either at rest or moving at a *constant* velocity, that is, at a constant speed in a fixed direction.

> If you are in a train or lift moving at a steady speed you cannot tell that you are not at rest – you are in equilibrium. When you speed up, slow down or go round a corner you do feel a force acting and you are no longer in equilibrium.

- Force is a vector quantity: it has both magnitude and direction.
- The unit of force is the newton, N.
- You specify a force by

either   or

giving its components, relative to a pair   giving its magnitude and direction, for
of basis vectors **i** and **j**   example, up-down, left-right, east-west
                                                or a bearing, say 090° (measured
                                                clockwise from north)

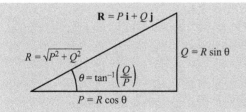

Finding a magnitude usually involves Pythagoras, theorem.

Finding the direction usually involves trigonometry.

- You can add forces by either adding the **i** and **j** components separately or by using a vector triangle.
- When a body is in equilibrium you can **resolve** the forces:

either   or

set the sum of all the forces to 0   set the sum of the forces in one direction
                                                equal to the sum of the forces in the
                                                opposite direction.

Example 1

A block of weight 24 N rests on two supports A and B

The reaction from B is twice as big as the force from A

a Sketch a force diagram.

b Work out the size of the reactions from A and B

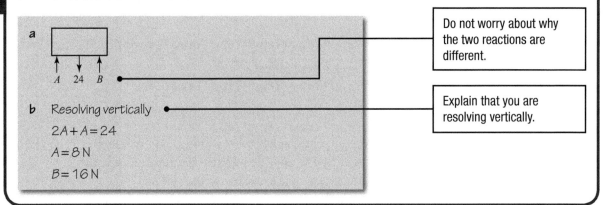

a

Do not worry about why the two reactions are different.

Explain that you are resolving vertically.

b Resolving vertically

$2A + A = 24$

$A = 8\,N$

$B = 16\,N$

Example 2

A stone has four forces acting on it: 10 N from the north, 5 N from the east, 7 N from the south and 12 N from the west.

Calculate the resultant force.

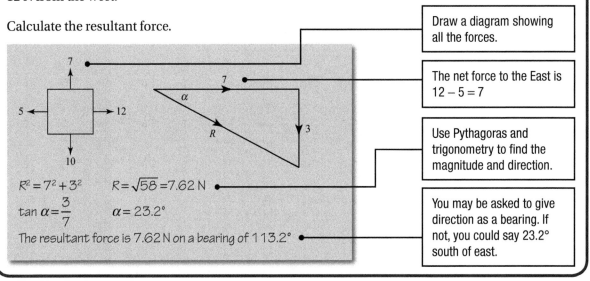

Draw a diagram showing all the forces.

The net force to the East is $12 - 5 = 7$

Use Pythagoras and trigonometry to find the magnitude and direction.

You may be asked to give direction as a bearing. If not, you could say 23.2° south of east.

$R^2 = 7^2 + 3^2$ $R = \sqrt{58} = 7.62\,N$

$\tan \alpha = \dfrac{3}{7}$ $\alpha = 23.2°$

The resultant force is 7.62 N on a bearing of 113.2°

Exam tips

- Look for key words that imply equilibrium; for example, at rest or constant velocity.
- Draw a large, clear diagram that shows all the forces acting on the body.
- If you want to show direction of motion ↑ or acceleration →→ draw the symbols separate from the body.
- Explain what you are doing; for example, say 'resolving forces horizontally'.
  You may do this using abbreviations, such as, R →.
- Remember to give units.
- Be prepared to identify any assumptions and say how changing the assumption might affect your answer.
  - a smooth surface   ignore friction
  - a rough surface   include friction
  - a body is a particle   ignore any effects of its size
  - a light object   ignore gravity for the object
  - a light string   tension is the same along its length
  - an inextensible string   ignore any stretching
  - a smooth pulley   tension is the same on each side of the pulley

1 Three forces act on a point particle: 10 N east, 15 N west and 25 N south.

What is the bearing of the resultant force? [3]

2 a Steve places his briefcase on a horizontal table.

Complete the diagram to show the forces acting on the briefcase. [1]

Briefcase

b A cat pushes the briefcase horizontally but the briefcase does not move.
Complete the diagram to show the forces acting on the briefcase. [2]

Briefcase

3 Calculate the resultant force in these situations.

Identify any that are in equilibrium.

a [1]

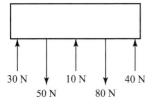

30 N   10 N   40 N
    50 N   80 N

3 b [2]

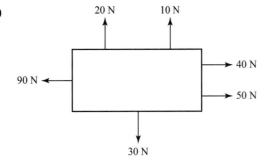

_____

_____

_____

_____

_____

c [4]

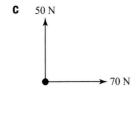

_____

_____

_____

_____

_____

_____

_____

_____

_____

_____

_____

4 Three forces $(5\mathbf{i} - 2\mathbf{j})$, $(3\mathbf{i} - \mathbf{j})$ and $(6\mathbf{i} + 7\mathbf{j})$ act on a body.

Calculate the magnitude and direction of the resultant force. [5]

5 Three forces act on a body $(6\mathbf{i} + 5\mathbf{j})$, $(-10\mathbf{i} - 13\mathbf{j})$ and $(a\mathbf{i} + b\mathbf{j})$

Find the integer values of a and b that give the smallest possible resultant force that acts in a direction 45° below the x-axis. [5]

6

You are given that this body is in equilibrium and that forces X and Y are positive.

Find the value of X and the value of Y [4]

Recap

- **Dynamics** is the study of forces and how they affect motion.
- **Newton's second law** is the most important rule used in dynamics.
  It connects the resultant force acting on a body to the acceleration of the body.

Newton's second law

A total force of $\mathbf{F}$ N acting on a body of mass m kg gives it an acceleration $\mathbf{a}$ m s^{-2} where

$$\mathbf{F} = m\mathbf{a}$$

- The force that appears in Newton's second law is the total force acting on the body.
  - Force must be measured in newtons, N
  - Mass must be measured in kilograms, kg
  - Acceleration must be measured in metres per second squared, m s^{-2}

Forces can be given in component form.

A force $\mathbf{F} = (10\mathbf{i} + 25\mathbf{j})$ N acting on a particle of mass 10 kg gives it an acceleration $\mathbf{a} = (1\mathbf{i} + 2.5\mathbf{j})$ m s^{-2}

- The force that appears in Newton's second law is the total force acting on the body.
- Given a constant acceleration you can apply the *suvat* equations, see section **7.3**, to calculate velocities, distances and times.

The *suvat* equations

$$v = u + at \qquad\qquad v^2 = u^2 + 2as$$
$$s = ut + \frac{1}{2}at^2$$
$$= vt - \frac{1}{2}at^2 \qquad\qquad s = \frac{1}{2}(u + v)t$$

- The *most important* part of an answer in a dynamics problem is likely to be a diagram showing all the forces acting on a body.
  - Make space for a large diagram (start a new page if necessary).
  - Do not rush this part of the answer.
  - Show all forces, including:
    a body's weight – the force due to gravity
    friction on rough surfaces
    normal reactions where bodies touch
    driving forces – due to engines
  - Show any acceleration using a double arrow →→

For a box being dragged along a rough surface

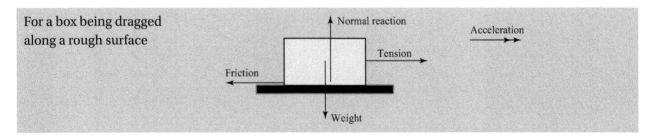

Example 1

The driving force acting on a car of mass 900 kg is 5000 N and the total resistive forces are 2000 N.

Calculate the acceleration of the car.

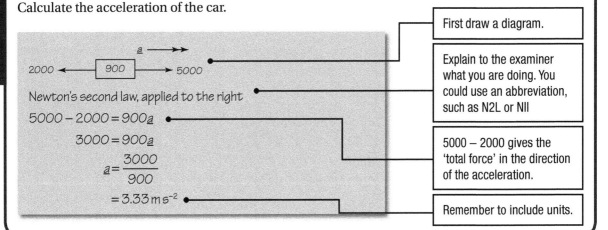

First draw a diagram.

Explain to the examiner what you are doing. You could use an abbreviation, such as N2L or NII

5000 – 2000 gives the 'total force' in the direction of the acceleration.

Remember to include units.

Example 2

A car, of mass 750 kg, is 60 m from some traffic lights, and it's travelling at 50 km h⁻¹ when the lights change from green to red.
Show that if the driver applies an average braking force of 1200 N, the car will *not* stop before the lights. Assume the braking force is the only force acting on the car.

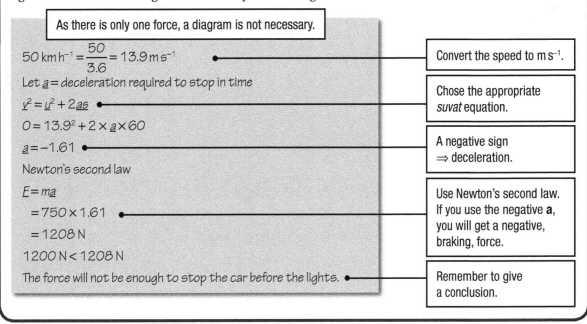

Convert the speed to m s⁻¹.

Chose the appropriate *suvat* equation.

A negative sign ⇒ deceleration.

Use Newton's second law. If you use the negative **a**, you will get a negative, braking, force.

Remember to give a conclusion.

Exam tips

- Start by drawing a large, clear diagram.
- Always tell the examiner what you are doing, for example, by writing 'using Newton's second law' or an abbreviation, such as, 'N2L.'
- Remember that the **F** in **F** = *m***a** means *total* force.
- A negative acceleration, or deceleration, simply means the velocity is decreasing in your chosen direction.
- Remember to use SI units, including in your answer.

1   A particle of mass 2 kg is acted on by a force of 12 N

What is the acceleration of the particle?   [2]

2   A crate of mass 100 kg is pushed across a smooth surface with a force of 20 N

a   Calculate the velocity of the crate after 15 seconds if it initially starts from rest.   [3]

b   After another 50 seconds the crate slides onto a rough surface and there is now a frictional force of 16 N.

After another 15 seconds the pushing force of 20 N is removed and the crate eventually stops moving.

What is the total length of time that the crate will have been in motion?   [8]

(*continued on next page*)

2 **b** (*continued*)

3 A metal sphere of mass 2 kg and weight 20 newtons is suspended from the roof of a train by a light inextensible string.

The train is accelerating at $1 \, \text{m s}^{-2}$

What angle does the string make with the vertical? [6]

4 **a** A lift is moving upwards and decelerating at 2 m s^{-2}
A woman of weight 600 N is standing in the lift.

What is the contact force from the floor on the woman? Use $g = 9.8 \text{ m s}^{-2}$ [4]

---

b Repeat part **a** if the lift is accelerating upwards at 2 m s^{-2} [2]

Recap

- The **mass** of a body is a property of the body. It is the same everywhere.
- Mass is measured in kilograms, kg, or related units such as milligrams, grams and tonnes.

    1000 milligrams = 1 gram

    1000 grams   = 1 kilogram

    1000 kilograms   = 1 tonne

- **Weight** is the *force* that acts on a massive body due to gravity.
  - Using Newton's second law, $\mathbf{F} = m\mathbf{a}$, the mass and weight of a body are related by the acceleration due to gravity.
  - To calculate the weight of a body multiply its mass by the acceleration due to gravity, g: $\mathbf{W} = mg$
- Acceleration due to gravity varies depending on position.

Mass 2kg

Weight = mass × acceleration due to gravity

$$\mathbf{W} = m\mathbf{g}$$
$$= 2 \times 9.81$$
$$= 19.6 \text{ N}$$

In the Arctic	$g = 9.83 \text{ m s}^{-2}$	1 kg weighs 9.83 N
In Peru	$g = 9.76 \text{ m s}^{-2}$	1 kg weighs 9.76 N
On the Moon	$g = 1.62 \text{ m s}^{-2}$	1 kg weighs 1.62 N
(Lunar gravity $\approx \frac{1}{6}$ Earth gravity)		
In free fall	$g = 0 \text{ m s}^{-2}$	1 kg weighs nothing

- In questions (because distances are small) you should take g to be a constant.
  - The average value of g on the Earth is 9.8 m s^{-2}, to 2 significant figures, and this is the value you should use unless told otherwise.
  - Exam papers may suggest you use 9.8, 9.81 or even 10

In questions involving weight you should expect to have to use other knowledge.

Newton's second law, $\mathbf{F} = m\mathbf{a}$, will allow you to calculate any acceleration.

The *suvat* equations will allow you to calculate subsequent distances, times and velocities.

In Newton's theory of universal gravity the force, F, between two bodies of masses, m and M, a distance, r, apart is given by

$$\mathbf{F} = G\frac{mM}{r^2}$$

$$\mathbf{W} = m \times \left(G\frac{M}{r^2} \right) \equiv m \times g$$

where G is the gravitational constant.

Clearly the acceleration due to gravity, g, depends on the mass of the body doing the attracting, M – gravity is weaker on the Moon than on Earth – and on the distance between the bodies, r – gravity is slightly weaker at the top of a mountain than at sea level.

Example 1

A farmer is ploughing a horizontal field.

The plough has a mass of 550 kg and is being pulled with a force of 800 N.

Given that the plough accelerates at 1.2 m s^{-2} and that **g** = 9.8 m s^2 calculate

a   The normal reaction from the field on the plough,

b   The total resistance to motion of the plough.

> In these examples, 'NI' and 'NII' are being used to refer to Newton's first and second laws – see sections 8.1 and 8.2 for a reminder.

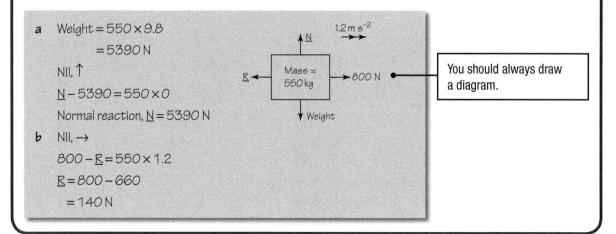

a   Weight = 550×9.8

   = 5390 N

NII, ↑

$\underline{N} - 5390 = 550 \times 0$

Normal reaction, $\underline{N} = 5390$ N

b   NII, →

$800 - \underline{R} = 550 \times 1.2$

$\underline{R} = 800 - 660$

   = 140 N

> You should always draw a diagram.

Example 2

A man of mass 70 kg stands on bathroom scales in an elevator that is moving upwards.
The scales show a reading of 89.5 kg.

a   Taking g as 9.81 m s^{-2}, calculate the acceleration of the lift.

b   Is the lift accelerating or decelerating?

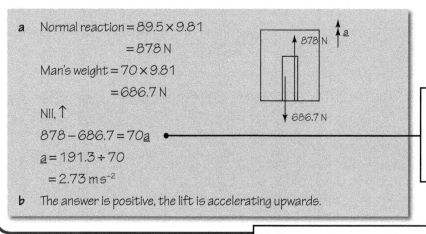

a   Normal reaction = 89.5×9.81

   = 878 N

Man's weight = 70×9.81

   = 686.7 N

NII, ↑

$878 - 686.7 = 70\underline{a}$

$\underline{a} = 191.3 \div 70$

   = 2.73 m s^{-2}

b   The answer is positive, the lift is accelerating upwards.

> Assume the lift is accelerating upwards. If this assumption is wrong, the answer for **a** will be negative.

> Bathroom scales measure the normal reaction they apply to a body. They represent this force as equivalent mass = normal reaction ÷ 9.81

Exam tips

- Read the question carefully: are you given a body's mass or weight?
  In everyday life people confuse the two ideas – you must be careful to distinguish between them!
- Remember that Newton's second law uses *mass* not weight.
- Remember to draw a large, clear diagram.
- If a body is falling freely under gravity its acceleration is 9.81 m s^{-2} and you should use this value in *suvat* calculations, unless the question tells you to use a different value.
  In practice, as a body accelerates downwards, air resistance increases and the acceleration reduces to 0 m s^{-2} when the air resistance exactly balances the body's weight. The body is said to have reached terminal velocity.

Use $g = 9.8$ m s^{-2} in this exercise.

1 The acceleration due to gravity on the planet Mars is 3.7 m s^{-2}. A spacecraft has mass 15 000 kg. What is its mass and weight on Mars? [1]

2 A woman has a weight of 430 N. Calculate her mass. [1]

3 A palette of 1000 bricks each of mass 2.8 kg is being raised off the ground by a rope at a constant velocity of 4 m s^{-1}.

a Calculate the tension in the rope. [2]

b How long will it take the palette of bricks to reach a platform that is 48 m above the ground? [1]

c Give two modelling assumptions that you have made about the rope. [2]

3 d A more accurate model of the bricks' movement assumes that they start from rest, accelerate uniformly at $1\,\mathrm{m\,s^{-2}}$ for the first 24 m and then decelerate at $1\,\mathrm{m\,s^{-2}}$ for the second 24 m.

    i Calculate the tension in the rope for the first 24 m. [2]

    ii Calculate the tension in the rope for the final 24 m. [1]

  e Calculate the length of time to raise the bricks to 48 m under this model. [4]

4 Jo has a mass of 81 kg. She stands on her bathroom scales in a lift.

What is the reading on the bathroom scales when the lift is

a moving at constant velocity, [1]

b accelerating upwards at $2\,\mathrm{m\,s^{-2}}$, [4]

c moving upwards and decelerating at $1\,\mathrm{m\,s^{-2}}$, [3]

d moving downwards and decelerating at $1\,\mathrm{m\,s^{-2}}$? [3]

Recap

- When bodies interact they exert forces on one another.
  - A massive body resting on a surface presses down with its weight.
    The surface pushes back with a **normal reaction**.
  - The Earth pulls a massive body downwards.
    The massive body pulls the Earth upwards.
  - A body being dragged across a rough surface feels friction, resisting its motion.
    The surface feels a force trying to drag it along with the body.
- Newton's third law says that these pairs of forces are **equal and opposite**.

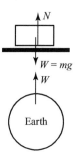

Newton's third law
If one body exerts a force **F** on a second body then the second body exerts an equal and opposite force –**F** on the first body.

- Bodies necessarily interact when they are connected. For example,
  - a car towing a trailer
  - masses connected by a light string over a frictionless pulley

The rod joining the car and trailer can be in thrust or tension.

When the car is braking, the thrust pushing the trailer back is equal to the thrust pushing the car forward.

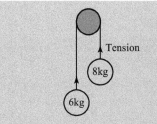

The string joining the masses can only be in tension.

The tension in the string is constant. The 6 kg weight is being pulled up by the same force as the 8 kg mass.

- The forces acting on connected bodies can either be
  - **external forces** such as weight or friction
    - You regard the Earth or a rough surface as an immovable object and ignore any forces acting on them.

  or

  - **internal forces** such as tension in a string or tension or thrust in a tow bar.
- You can either
  - consider the connected bodies as one single body
    - In this case you only consider the external forces.

By Newton's third law, since all internal forces come in equal and opposite pairs then they do not contribute to the resultant force on a system of connected bodies. The connected bodies can be considered as a single object with only the external forces acting on it.

  - consider the individual bodies separately
    - In this case you must consider both the internal and external forces.

  However, usually you will need to do both.

Example 1

A family car has mass 800 kg and is pulling a trailer of mass 200 kg.
It is accelerating at 4 m s^{-2}

The total resistance to motion for the car is 150 N and for the trailer is 250 N.

a Find the driving force of the car.

b Find the force in the tow bar and determine whether it is a thrust or a tension.

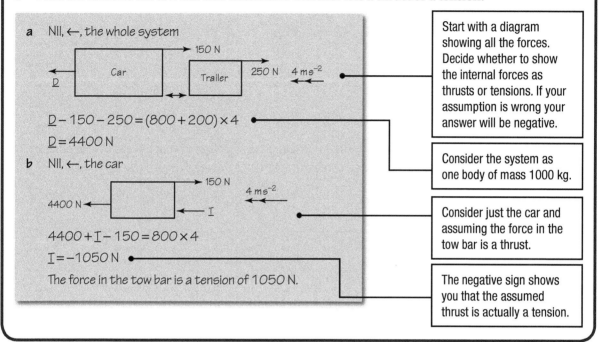

a NII, ←, the whole system

$$D - 150 - 250 = (800 + 200) \times 4$$
$$D = 4400\,N$$

b NII, ←, the car

$$4400 + \underline{T} - 150 = 800 \times 4$$
$$\underline{T} = -1050\,N$$

The force in the tow bar is a tension of 1050 N.

Start with a diagram showing all the forces. Decide whether to show the internal forces as thrusts or tensions. If your assumption is wrong your answer will be negative.

Consider the system as one body of mass 1000 kg.

Consider just the car and assuming the force in the tow bar is a thrust.

The negative sign shows you that the assumed thrust is actually a tension.

Example 2

A mass of 4 kg is held at rest 50 cm above the ground. It is attached, by a light, inextensible string that passes over a smooth, light pulley, to a mass of 2 kg that lies at rest on a smooth, horizontal surface. The 4 kg mass is then released from rest.

a Calculate the tension in the string.

b After what time will the 4 kg mass hit the ground.

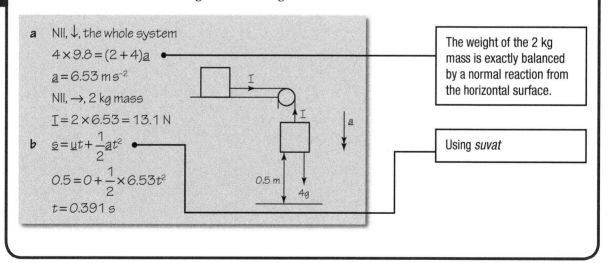

a NII, ↓, the whole system
$$4 \times 9.8 = (2 + 4)\underline{a}$$
$$\underline{a} = 6.53\,ms^{-2}$$
NII, →, 2 kg mass
$$\underline{T} = 2 \times 6.53 = 13.1\,N$$

b $s = \underline{u}t + \dfrac{1}{2}\underline{a}t^2$

$$0.5 = 0 + \frac{1}{2} \times 6.53t^2$$
$$t = 0.391\,s$$

The weight of the 2 kg mass is exactly balanced by a normal reaction from the horizontal surface.

Using *suvat*

Exam tips

- Draw a clear diagram showing all the forces.
- Be aware of which forces are external and which internal.
- Usually consider the whole system first (ignoring internal forces).
- Remember that a string can only be in tension, whereas a bar can be in tension or thrust.

1 A lift is moving downwards at constant speed.

Calculate the normal reaction from the floor of the lift on a man of mass 80 kg. [2]

2 A 9 kg mass is held at rest and is connected to a 4 kg mass by a string that passes over a pulley.
A 3 kg mass hangs below the 4 kg mass also connected by a shorter length of string.

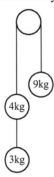

a Calculate the acceleration of the system after the 9 kg mass is released. [4]

b Show that the tension in the longer string is 77.2 N [3]

2   c Calculate the tension in the shorter string. [2]

---

d Write down two assumptions you have made about the pulley and two assumptions you have made about the string. [2]

---

3 A train consists of a miniature locomotive pulling two carriages.
The mass of the locomotive is 600 kg and the mass of each carriage is 800 kg.
The resistance to motion of the locomotive is 100 N and each carriage has resistance 300 N.

a The train accelerates from rest at $1.5 \, \mathrm{m \, s^{-2}}$

Calculate the driving force of the train. [3]

---

3 b The couplings between the carriages and between the locomotive and the carriage are all rigid.

Calculate the force in the coupling between the carriages.
Is it a tension or a thrust?   [3]

----

c The maximum speed of the train is $5\,\mathrm{m\,s^{-1}}$

Calculate the braking force needed to bring the train to rest in 10 metres when travelling
at full speed.   [5]

Recap

- As long as the x- and y-components of acceleration are constant, the equations of motion for two-dimensional motion are vector versions of four of the five *suvat* equations for straight line motion.

Straight line motion	Two-dimensional motion
$v = u + at$	$\mathbf{v} = \mathbf{u} + \mathbf{a}t$
$s = \dfrac{1}{2}(u+v)t$	$\mathbf{s} = \dfrac{1}{2}(\mathbf{u}+\mathbf{v})t$
$s = ut + \dfrac{1}{2}at^2$	$\mathbf{s} = \mathbf{u}t + \dfrac{1}{2}\mathbf{a}t^2$
$s = vt - \dfrac{1}{2}at^2$	$\mathbf{s} = \mathbf{v}t - \dfrac{1}{2}\mathbf{a}t^2$

- Initial velocity can be written $\mathbf{u} = u_x\mathbf{i} + u_y\mathbf{j}$
- Final velocity can be written $\mathbf{v} = v_x\mathbf{i} + v_y\mathbf{j}$
- Acceleration can be written $\mathbf{a} = a_x\mathbf{i} + a_y\mathbf{j}$
- The vector version of $v^2 = u^2 + 2as$ is beyond the scope of this book, but the equation can still be applied to the individual components:

$$v_x^2 = u_x^2 + 2a_x x \qquad v_y^2 = u_y^2 + 2a_y y$$

- The vector equations work because the non-vector equations can be applied independently to each component.
- Vectors can be written using

  - unit vectors

    For example a velocity of $(6\mathbf{i} - 2\mathbf{j})\,\text{m s}^{-1}$

  - column vectors

    For example an acceleration of $\begin{pmatrix} 0 \\ -9.81 \end{pmatrix}\text{m s}^{-2}$

- Be careful not to confuse displacement, $\mathbf{s}$, with the position of the particle, often labelled $\mathbf{r}$

  - The position vector of point (x_1, y_1) is $\begin{pmatrix} x_1 \\ y_1 \end{pmatrix}$

  - If a particle moves from (x_1, y_1) to (x_2, y_2) its displacement is $\begin{pmatrix} x_2 \\ y_2 \end{pmatrix} - \begin{pmatrix} x_1 \\ y_1 \end{pmatrix} = \begin{pmatrix} x_2 - x_1 \\ y_2 - y_1 \end{pmatrix}$

    If a particle moves from $(-3, 9)$ to $(18, 20)$ its displacement is
    $$\begin{pmatrix} 18 \\ 20 \end{pmatrix} - \begin{pmatrix} -3 \\ 9 \end{pmatrix} = \begin{pmatrix} 18 - -3 \\ 20 - 9 \end{pmatrix} = \begin{pmatrix} 21 \\ 11 \end{pmatrix}$$

- When you write vectors by hand remember to underline them to show they are vectors.

Example 1

A particle starts accelerating from a point with position vector $(4\mathbf{i} + 10\mathbf{j})\,\text{m}$ and finishes 8 seconds later at the point with position vector $(8\mathbf{i} - 6\mathbf{j})\,\text{m}$.

During this time its velocity increases to three times its initial value.

Find the initial speed of the particle.

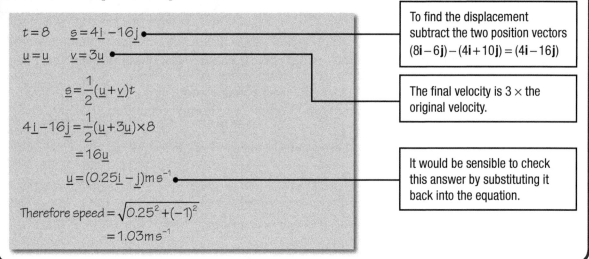

$t = 8 \qquad \underline{s} = 4\underline{i} - 16\underline{j}$

$\underline{u} = \underline{u} \qquad \underline{v} = 3\underline{u}$

$\underline{s} = \dfrac{1}{2}(\underline{u} + \underline{v})t$

$4\underline{i} - 16\underline{j} = \dfrac{1}{2}(\underline{u} + 3\underline{u}) \times 8$

$= 16\underline{u}$

$\underline{u} = (0.25\underline{i} - \underline{j})\,\text{m s}^{-1}$

Therefore speed $= \sqrt{0.25^2 + (-1)^2}$

$= 1.03\,\text{m s}^{-1}$

To find the displacement subtract the two position vectors $(8\mathbf{i} - 6\mathbf{j}) - (4\mathbf{i} + 10\mathbf{j}) = (4\mathbf{i} - 16\mathbf{j})$

The final velocity is $3 \times$ the original velocity.

It would be sensible to check this answer by substituting it back into the equation.

Example 2

A particle has initial velocity $(6\mathbf{i} + 3\mathbf{j})\,\text{m s}^{-1}$ and accelerates at $(\mathbf{i} - 3\mathbf{j})\,\text{m s}^{-2}$

Its final velocity has no x-component and its speed is unchanged.

Find the displacement of the particle during this motion.

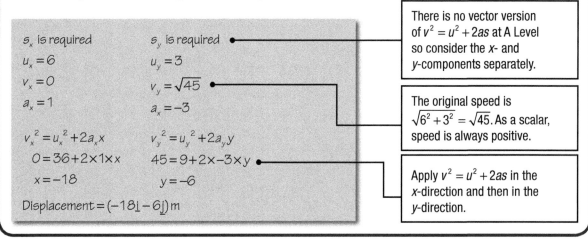

s_x is required

$u_x = 6$

$v_x = 0$

$a_x = 1$

$v_x^2 = u_x^2 + 2a_x x$

$0 = 36 + 2 \times 1 \times x$

$x = -18$

s_y is required

$u_y = 3$

$v_y = \sqrt{45}$

$a_x = -3$

$v_y^2 = u_y^2 + 2a_y y$

$45 = 9 + 2 \times -3 \times y$

$y = -6$

Displacement $= (-18\underline{i} - 6\underline{j})\,\text{m}$

There is no vector version of $v^2 = u^2 + 2as$ at A Level so consider the x- and y-components separately.

The original speed is $\sqrt{6^2 + 3^2} = \sqrt{45}$. As a scalar, speed is always positive.

Apply $v^2 = u^2 + 2as$ in the x-direction and then in the y-direction.

Exam tips

- Start by listing the *suvat* values/vectors you are given in the question and write down what you need to find out.
- To find the final position of an object after a motion, use *suvat* to find the displacement, but remember to add on any initial displacement to find the final position vector relative to the origin.
- Check whether you are asked to give a vector (velocity/displacement) or a scalar (speed/distance) for your answer.

1   A particle has initial velocity $(5\mathbf{i} + 6\mathbf{j})\,\mathrm{m\,s^{-1}}$. After 8 seconds its velocity is $3\mathbf{i} - 2\mathbf{j}$

What is the acceleration of the particle, in $\mathrm{m\,s^{-2}}$?   [2]

2   A particle moves from $(3, 4)$ to $(12, 10)$ with acceleration $(6\mathbf{i} - 2\mathbf{j})\,\mathrm{m\,s^{-2}}$ for 10 seconds.

What is the final velocity of the particle?   [4]

3 A particle has initial velocity $(10\mathbf{i} + 6\mathbf{j})\,\mathrm{m\,s^{-1}}$ and accelerates to $(12\mathbf{i} - 4\mathbf{j})\,\mathrm{m\,s^{-1}}$ whilst moving from $(-2, 5)$ to $(8, 13)$

Find the acceleration of the particle. [6]

4 It is given that $\mathbf{i}$ and $\mathbf{j}$ are unit vectors in the east and north directions respectively.
A particle starts moving with velocity $(5\mathbf{i} - 32\mathbf{j})\,\mathrm{m\,s^{-1}}$ and accelerates at a constant $(4\mathbf{i} - \mathbf{j})\,\mathrm{m\,s^{-2}}$
After how many seconds is the particle moving south-east? [3]

Motion in two dimensions Two-dimensional motion with constant acceleration

5 A particle starts with velocity $(4\mathbf{i} + 6\mathbf{j})\,\text{m s}^{-1}$ and accelerates at $(4\mathbf{i} - 7\mathbf{j})\,\text{m s}^{-2}$ for 2 seconds.

Show that the speed of the particle doubles. [6]

6 A particle has an initial velocity of $(6\mathbf{i} - 2\mathbf{j})\,\text{m s}^{-1}$ and accelerates for 4 seconds so that its direction of travel is perpendicular to the original direction but its speed is unchanged.

Find the two possible values of the acceleration. [6]

Recap

- This diagram should be familiar from section **7.4**. It summarises how to use calculus to move between displacement, velocity and acceleration for motion in a straight line.

  Exactly the same relationships hold for motion in two dimensions but the calculus is applied to **vectors** rather than scalars.

- If **s** is displacement, **v** is velocity and **a** is acceleration where **s**, **v** and **a** are 2D vectors

  - $\mathbf{v} = \dfrac{d\mathbf{s}}{dt} = \dot{\mathbf{s}}$

  - $\mathbf{a} = \dfrac{d\mathbf{v}}{dt} = \dot{\mathbf{v}} = \dfrac{d^2\mathbf{s}}{dt^2} = \ddot{\mathbf{s}}$

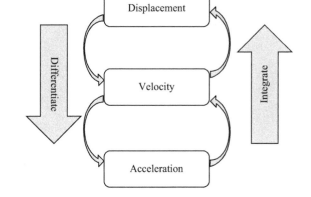

- To differentiate a vector, you differentiate the components.
- To integrate a vector, you integrate the components.
  The constant of integration is also a vector.

If	$\mathbf{s} = (3t-5)\mathbf{i} - t^2\mathbf{j}$	If	$\mathbf{a} = \sin 2t\,\mathbf{i} + \cos t\,\mathbf{j}$
then	$\begin{aligned}\mathbf{v} &= \dfrac{d\mathbf{s}}{dt} \\ &= \dfrac{d}{dt}(3t-5)\mathbf{i} - \dfrac{d}{dt}t^2\mathbf{j} \\ &= 3\mathbf{i} - 2t\mathbf{j}\end{aligned}$	then	$\begin{aligned}\mathbf{v} &= \int \mathbf{a}\,dt \\ &= \int \sin 2t\,dt\,\mathbf{i} + \int \cos t\,dt\,\mathbf{j} \\ &= -\tfrac{1}{2}\cos 2t\,\mathbf{i} + \sin t\,\mathbf{j} + \mathbf{c}\end{aligned}$

Example 1

A particle has displacement $\mathbf{s} = (3t\mathbf{i} + t^2\mathbf{j})\,\text{m}$

Find the speed of the particle after 4 seconds.

$\underline{s} = (3t\underline{i} + t^2\underline{j})$

$\underline{v} = 3\underline{i} + 2t\underline{j}$ ●———————— Differentiate the components of **s** to get the components of **v**

So after 4 s, $\underline{v} = 3\underline{i} + 8\underline{j}$

$\text{Speed} = \sqrt{3^2 + 8^2}$ ●———————— Use Pythagoras' theorem as speed is the magnitude of the velocity.

$\qquad = 8.54\,\text{m s}^{-1}$

Example 2

A particle starts from the point $(6, -2)$ moving with speed $26\,\text{m s}^{-1}$ at an angle of $\tan^{-1}\left(\dfrac{12}{5}\right)$ above the x-axis and accelerating at $(6t\mathbf{i} - 4\mathbf{j})\,\text{m s}^{-2}$

Find the position of the particle after 3 seconds.

initial velocity $= 26\cos\theta\,\underline{\mathbf{i}} + 26\sin\theta\,\underline{\mathbf{j}}$

$\qquad\qquad\quad = 10\underline{\mathbf{i}} + 24\underline{\mathbf{j}}$

$\underline{\mathbf{a}} = 6t\underline{\mathbf{i}} - 4\underline{\mathbf{j}}$

$\underline{\mathbf{v}} = 3t^2\underline{\mathbf{i}} - 4t\underline{\mathbf{j}} + \mathbf{c}$

When $t = 0,\ \underline{\mathbf{v}} = 10\underline{\mathbf{i}} + 24\underline{\mathbf{j}}$

$10\underline{\mathbf{i}} + 24\underline{\mathbf{j}} = 0\underline{\mathbf{i}} + 0\underline{\mathbf{j}} + \mathbf{c}$

$\underline{\mathbf{v}} = (3t^2 + 10)\underline{\mathbf{i}} + (-4t + 24)\underline{\mathbf{j}}$

$\underline{\mathbf{s}} = (t^3 + 10t)\underline{\mathbf{i}} + (-2t^2 + 24t)\underline{\mathbf{j}} + \mathbf{c}_2$

When $t = 0,\ \underline{\mathbf{s}} = 6\underline{\mathbf{i}} - 2\underline{\mathbf{j}}$

$6\underline{\mathbf{i}} - 2\underline{\mathbf{j}} = 0\underline{\mathbf{i}} + 0\underline{\mathbf{j}} + \mathbf{c}_2$

$\underline{\mathbf{s}} = (t^3 + 10t + 6)\underline{\mathbf{i}} + (-2t^2 + 24t - 2)\underline{\mathbf{j}}$

Therefore when $t = 3,\ \underline{\mathbf{s}} = 63\underline{\mathbf{i}} + 52\underline{\mathbf{j}}$

Write the initial velocity in component form, by sketching a triangle using θ as the angle with the x-axis.

Pythagoras' theorem gives the hypotenuse as 13 so $\sin\theta = \dfrac{12}{13}$ and $\cos\theta = \dfrac{5}{13}$

Integrate **a** to get **v**. Remember that the constant is a vector.

Use the initial **v** value to find **c**

Integrate **v** to get **s**

When $t = 0$, $\mathbf{s} = 6\mathbf{i} - 2\mathbf{j}$

Example 3

A particle of mass $3\,\text{kg}$ has initial velocity $(18\mathbf{i} - 18\mathbf{j})\,\text{m s}^{-1}$ and is acted on by a force of $12t\mathbf{i} + 9\mathbf{j}\,\text{N}$

Find its speed when it is travelling parallel to the x-axis.

$\qquad\quad \underline{\mathbf{F}} = m\underline{\mathbf{a}}$

$12t\underline{\mathbf{i}} + 9\underline{\mathbf{j}} = 3\underline{\mathbf{a}}$

$\qquad\quad \underline{\mathbf{a}} = 4t\underline{\mathbf{i}} + 3\underline{\mathbf{j}}$

$\qquad\quad \underline{\mathbf{v}} = 2t^2\underline{\mathbf{i}} + 3t\underline{\mathbf{j}} + \underline{\mathbf{c}}$

When $t = 0,\ v = 18\underline{\mathbf{i}} - 18\underline{\mathbf{j}}$

Therefore $\quad \underline{\mathbf{v}} = 2t^2\underline{\mathbf{i}} + 3t\underline{\mathbf{j}} + 18\underline{\mathbf{i}} - 18\underline{\mathbf{j}}$

$\qquad\qquad\quad = (2t^2 + 18)\underline{\mathbf{i}} + (3t - 18)\underline{\mathbf{j}}$

$3t - 18 = 0$

$\qquad t = 6$

$\underline{\mathbf{v}} = (2 \times 6^2 + 18)\underline{\mathbf{i}} + 0\underline{\mathbf{j}}$

$\quad = 90\underline{\mathbf{i}}$

Therefore speed $= 90\,\text{m s}^{-1}$

Using Newton's 2nd law.

Integrate **a** to get **v**
Don't forget the vector constant.

Use the given value of **v** to find the constant.

When travelling parallel to the x-axis the component of $\mathbf{j} = 0$

Exam tips

- If you see expressions for **s**, **v** or **a** in terms of t, this is an indication that you should use calculus and not *suvat*.
- Remember the constant when integrating and that it is a **vector**.

1 A particle has velocity $(5t^2\mathbf{i} - 2t\mathbf{j})\,\mathrm{m\,s^{-1}}$

   What is its acceleration when $t = 4$? [3]

2 A particle has displacement $\left((t^3 - 6t + 3)\mathbf{i} + \sqrt{t}\,\mathbf{j}\right)\mathrm{m}$

   Find its acceleration when $t = 9$ [3]

3 A particle starts from rest at the point $(4, -3)$ with acceleration $(2\mathbf{i} - 3\mathbf{j})\,\mathrm{m\,s^{-2}}$

   Find its position when $t = 2$ [4]

4 A particle starts at the origin with a velocity of $(2\mathbf{i} + 0.4\mathbf{j})$ and accelerates at $(8e^{4t}\mathbf{i} + 0.8e^{2t}\mathbf{j})\,\text{m s}^{-2}$ where $\mathbf{i}$ and $\mathbf{j}$ are unit vectors in the directions east and north respectively.

a Find an expression for the displacement of the particle at time t [5]

b Show that there is no time when the particle is north–east of the origin. [4]

5 A particle has displacement $\mathbf{s}$ m where $\mathbf{s} = \sin 2t\,\mathbf{i} + \cos 2t\,\mathbf{j}$

   a Given that its acceleration at time t is given by $\mathbf{a}$ m s^{-2}, show that $\mathbf{a} = -4\mathbf{s}$ [3]

   b Find the magnitude and direction of the initial velocity.   [2]

6 A particle moves such that its velocity at time t s is given by $(\mathbf{i} + \ln t\,\mathbf{j})$ m s^{-1}

   Given that the particle has displacement $(e^2\mathbf{i} + e^2\mathbf{j})$ m when $t = e^2$ s find an expression for the displacement of the particle at time t s   [4]

Recap

- When an object moves as a projectile, its horizontal component of velocity stays constant but its vertical component accelerates under gravity. Therefore, as a vector, the acceleration of a projectile is $\mathbf{a} = -g\mathbf{j}$

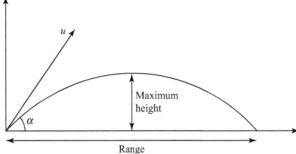

- If a particle is projected with initial speed u at angle α to the horizontal, its initial velocity as a vector is $\mathbf{v} = u\cos\alpha\,\mathbf{i} + u\sin\alpha\,\mathbf{j}$

- Integrating $\mathbf{a} = -g\mathbf{j}$ gives $\mathbf{v} = -gt\mathbf{j} + \mathbf{c}$ and using the initial velocity gives $\mathbf{v} = u\cos\alpha\,\mathbf{i} + (u\sin\alpha - gt)\mathbf{j}$

- Integrating again gives $\mathbf{s} = ut\cos\alpha\,\mathbf{i} + (ut\sin\alpha - \frac{1}{2}gt^2)\mathbf{j} + \mathbf{c}_2$, but if we define the origin as the point of projection, $\mathbf{c}_2 = 0$ so $\mathbf{s} = ut\cos\alpha\,\mathbf{i} + (ut\sin\alpha - \frac{1}{2}gt^2)\mathbf{j}$

- This formula can be used to derive equations for range and maximum height of a projectile. You should learn how to derive these.

Range

At maximum range, $s_y = 0$

$$ut\sin\alpha - \frac{1}{2}gt^2 = 0 \implies t = 0$$

$$\text{or } t = \frac{2u\sin\alpha}{g}$$

Then substitute this value for t into the horizontal component of displacement, $s_x = ut\cos\alpha$

$$\text{Range} = \frac{u^2\sin 2\alpha}{g}$$

$[\sin 2\alpha = 2\sin\alpha\cos\alpha]$

Maximum height

At maximum height $v_y = 0$

$$u\sin\alpha - gt = 0 \implies t = \frac{u\sin\alpha}{g}$$

Then substitute $v = 0$ into the formula $v^2 = u^2 + 2as$

$$\text{Maximum height} = \frac{u^2\sin^2\alpha}{2g}$$

- The path (or trajectory) of a projectile is a parabola. The equation of the path connects the horizontal and vertical components, x and y, of a general point on the curve after t seconds. You do not need to learn the equation of the path, but you should be able to derive it.

Equation of the path

Write down the horizontal and vertical components of displacement and eliminate t

Horizontally $x = ut\cos\alpha \implies t = \dfrac{x}{u\cos\alpha}$

Vertically $y = ut\sin\alpha - \dfrac{1}{2}gt^2$

Substitute for t $y = u\dfrac{x}{u\cos\alpha}\sin\alpha - \dfrac{1}{2}g\left(\dfrac{x}{u\cos\alpha}\right)^2$

$$= x\tan\alpha - \frac{gx^2}{2u^2\cos^2\alpha}$$

Use $\dfrac{1}{\cos\alpha} = \sec\alpha$ $y = x\tan\alpha - \dfrac{g\sec^2\alpha}{2u^2}x^2$

A particle is projected from the origin with initial velocity u m s^{-1} at $\theta°$ above the horizontal. x is horizontal and y is vertical. The equation of its path is

$$y = x\tan\theta - \frac{g\sec^2\theta}{2u^2}x^2$$

Given that $u = 7$ and the particle passes through the point $(\sqrt{20}, 0.5)$, find the value of θ

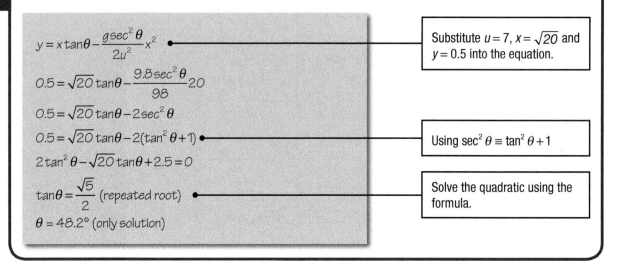

$$y = x\tan\theta - \frac{g\sec^2\theta}{2u^2}x^2$$

Substitute $u = 7$, $x = \sqrt{20}$ and $y = 0.5$ into the equation.

$$0.5 = \sqrt{20}\tan\theta - \frac{9.8\sec^2\theta}{98}20$$

$$0.5 = \sqrt{20}\tan\theta - 2\sec^2\theta$$

$$0.5 = \sqrt{20}\tan\theta - 2(\tan^2\theta + 1)$$

Using $\sec^2\theta \equiv \tan^2\theta + 1$

$$2\tan^2\theta - \sqrt{20}\tan\theta + 2.5 = 0$$

$$\tan\theta = \frac{\sqrt{5}}{2} \text{ (repeated root)}$$

Solve the quadratic using the formula.

$$\theta = 48.2° \text{ (only solution)}$$

A tennis player returns a ball from 0.8 m above the base line with a speed of 28 m s^{-1} at 5° above the horizontal.

The net in the centre of the court is 0.914 m high and the whole court is 23.77 m long.

a Show that the ball clears the net.

b Write down two assumptions made in your calculation.

Explain that 11.885 is half the length of the court.

a $11.885 = 28\cos5°t$

Using $s = ut$ for horizontal motion and the net is halfway along the court.

$t = 0.426$ s

$s = 28\sin5 \times 0.426 - 4.9 \times 0.426^2$

Using $s = ut + \frac{1}{2}at^2$ vertically.

$= 0.150$

Height of the ball $= 0.8 + 0.150 = 0.950$

The player clears the net by
$0.950 - 0.914 = 0.036\,m = 3.6\,cm$

State your conclusion clearly.

b Assuming the ball
- is a particle
- does not spin
- is light
- is not affected by air resistance
- moves only in 2D (i.e. not side to side)

Any two of these would suffice.

There could be other valid assumptions.

Exam tips

- Remember to consider the components separately:
  - vertically using *suvat* with $a = -g$
  - horizontally with constant velocity.
- You can use symmetry:
  - time to maximum height is half the total time.
  - the horizontal distance to maximum height is half the range.
- If you derive an equation as the first part of a question expect a later part to use that equation.
- Be prepared to describe clearly any modelling assumptions.

Where necessary take g as $9.8\,\text{m}\,\text{s}^{-2}$ unless otherwise stated.

1 A particle is projected over level ground with initial speed $24\,\text{m}\,\text{s}^{-1}$ at $30°$ above the horizontal.

Show that it is above a height of $3\,\text{m}$ for less than $2\,\text{s}$ [4]

2 a Show that the maximum height of a projectile with initial velocity u at $\alpha°$ to the horizontal

is $\dfrac{u^2 \sin^2 \alpha}{2g}$ [3]

b A particle projected at $8 \, \text{m s}^{-1}$ at $60°$ to the horizontal rises to a maximum height of $2.64 \, \text{m}$.

What value does this suggest for g? [2]

_____

_____

_____

_____

_____

3 A particle is projected over level ground with initial speed $10 \, \text{m s}^{-1}$ at $\alpha°$ above the horizontal. The particle hits the ground $5 \, \text{m}$ from the point of projection.

Taking g as $10 \, \text{m s}^{-2}$, find the two possible values of α [6]

_____

_____

_____

_____

_____

_____

_____

_____

_____

_____

_____

_____

_____

_____

4 A projectile with initial velocity u at $\alpha°$ to the horizontal passes through the point with coordinates (x, y) taking the point of projection as the origin.

a Show that $y = x\tan\alpha - \dfrac{g\sec^2\alpha}{2u^2}x^2$ [3]

b Given that $\alpha = 30°$ show that the particle moves with gradient $\dfrac{1}{5}$ when $x = \dfrac{u^2(5\sqrt{3}-3)}{20g}$ [6]

Recap

- When you add two forces acting at right angles, you use Pythagoras' theorem and trigonometry to find the magnitude and direction of the **resultant** force.

A 15 N force acts north and a 20 N force acts east.

The resultant magnitude, $R = \sqrt{15^2 + 20^2} = 25$ N

The direction of the resultant, $\alpha = \tan^{-1}\left(\dfrac{15}{20}\right) = 36.9°$

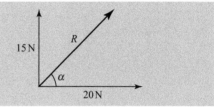

- The reverse of this process, to split a force into two **components** is called **resolving** the force.
- You resolve in two directions that are at right angles to each other, generally
  - the x- and y-directions,
  - horizontally and vertically or
  - parallel and perpendicular to an inclined plane.

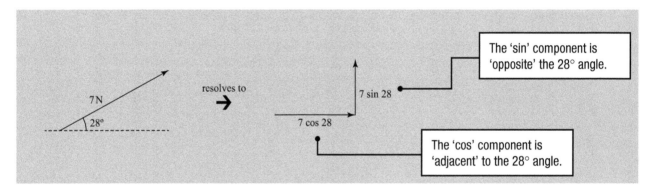

The 'sin' component is 'opposite' the 28° angle.

The 'cos' component is 'adjacent' to the 28° angle.

- If a force is given as unit vectors, for example $\mathbf{F} = 7\mathbf{i} - 2\mathbf{j}$, it is already in component form so these sort of problems can often be easier.
- If a particle is in **equilibrium** and therefore not accelerating then the forces will balance and you can use Newton's first law: the resultant force is zero.
- If a particle is not in equilibrium then it is accelerating and the resultant force is not zero. In this situation you can use Newton's second law: $F = ma$ or the equivalent vector version $\mathbf{F} = m\mathbf{a}$
- Remember that a particle may be in equilibrium in one direction but accelerating in the perpendicular direction.

An object of mass m kg on a smooth table is pulled by a string with tension T at an angle α

The object is in equilibrium perpendicular to the table so the forces balance: $\qquad R + T\sin\alpha = mg$

There is a resultant force parallel to the table so the object accelerates: $\qquad F = T\cos\alpha = ma$

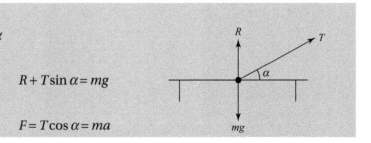

Example 1

A particle of mass 4 kg is acted on by three forces as shown in the diagram.

The particle accelerates northward.

a Find the acceleration.

b Find X

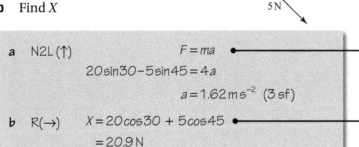

> As the acceleration is northward, use Newton's second law for the components in this direction.
> Note X does not have a component in this direction.

a N2L (↑) $F = ma$

$20\sin30 - 5\sin45 = 4a$

$a = 1.62\,\mathrm{ms^{-2}}$ (3 sf)

b R(→) $X = 20\cos30 + 5\cos45$

$= 20.9\,\mathrm{N}$

> There is no acceleration in the east–west direction so there is no resultant force.
> You could write
> $20\cos30° + 5\cos45° - X = 0$
> if you prefer.

Example 2

A particle of mass 5 kg rests on a rough slope inclined at 10° to the horizontal.

a Find the size of the frictional force acting up the slope. Take $g = 9.8\,\mathrm{m\,s^{-2}}$

The particle is now pulled up the slope by a string inclined at 10° to the slope with a force of 120 N The direction of the friction is changed but its magnitude remains the same.

b Find the acceleration of the particle.

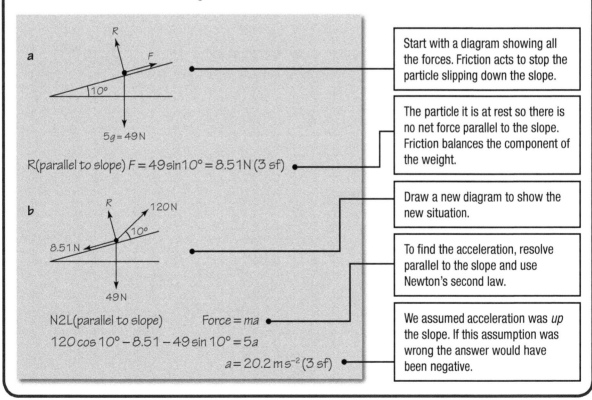

a

$5g = 49\,\mathrm{N}$

R(parallel to slope) $F = 49\sin10° = 8.51\,\mathrm{N}$ (3 sf)

> Start with a diagram showing all the forces. Friction acts to stop the particle slipping down the slope.

> The particle it is at rest so there is no net force parallel to the slope. Friction balances the component of the weight.

b

$8.51\,\mathrm{N}$ 120 N R 10°

$49\,\mathrm{N}$

N2L (parallel to slope)   Force $= ma$

$120\cos10° - 8.51 - 49\sin10° = 5a$

$a = 20.2\,\mathrm{ms^{-2}}$ (3 sf)

> Draw a new diagram to show the new situation.

> To find the acceleration, resolve parallel to the slope and use Newton's second law.

> We assumed acceleration was *up* the slope. If this assumption was wrong the answer would have been negative.

Exam tips

- Always draw a clearly labelled diagram to show the situation.
- Remember that friction is a force resisting motion and acts in a direction that directly opposes the motion.
- Remember that Newton's second law uses *mass* not weight and mass *is not* a force.

MyMaths   🔍 2192   SEARCH

Where necessary take g as $9.8\,\mathrm{m\,s^{-2}}$

1   A particle of mass 2 kg is acted on by forces of $(4\mathbf{i} - 6\mathbf{j})\,\mathrm{N}$, $(2\mathbf{i} + 3\mathbf{j})\,\mathrm{N}$ and $7\mathbf{j}\,\mathrm{N}$

   Find the magnitude and direction of the acceleration.   [5]

2   A particle of mass 5 kg is acted on by forces of 15 N, 12 N and 14 N on bearings of 340°, 040° and 190° respectively.

   a   Find the magnitude and the direction of the acceleration with respect to the $\mathbf{i}$ direction.   [6]

2 b A fourth force is added so that the particle now accelerates north.

Describe this force. [2]

_____

_____

_____

_____

3 A kite surfer of weight 750 N is being pulled by a rope angled at 32° to the horizontal with a tension of 1415 N
At time $t = 0$ the surfer is moving at $5\,\mathrm{m\,s^{-1}}$ and experiences a drag from the water of 500 N

a How far will the surfer have travelled when $t = 10\,\mathrm{s}$? [5]

_____

_____

_____

_____

_____

b What assumption has been made in your calculation?

How would your answer have been affected if this assumption had not been made? [3]

_____

_____

_____

_____

_____

4   A block of mass 5 kg is being pulled from rest along a rough horizontal surface by a string inclined at 45° to the horizontal. The frictional force is equal to half of the normal reaction. Given that the block is travelling at a speed of 5 m s⁻¹ after moving for 5 seconds, find the tension in the string.   [5]

5   A particle of mass 5 kg has initial velocity of $4\mathbf{i}$ m s⁻¹ and is acted on by forces of $(3\mathbf{i} + 5\mathbf{j})$ N, T N and $(8\mathbf{i} + 4\mathbf{j})$ N

Given that the velocity of the particle after 10 s is $(34\mathbf{i} + 16\mathbf{j})$ m s⁻¹, find T [4]

Recap

- Statics deals with bodies where the forces acting are in **equilibrium**. This means there is no overall net force so there is no acceleration. This happens when an object is stationary, is moving at constant velocity or is on the point of moving.
- When an object is at rest on a rough horizontal surface, it is in equilibrium. The weight, W, acting downwards is balanced by the normal reaction force, R, acting upwards.
- If a gentle push force, P, is applied parallel to the surface, **friction**, F, acts to oppose any motion of the object.

  F is equal and opposite to P.
- As P increases, F also increases up to a maximum value, F_{max} When the object is on the point of moving it is in **limiting equilibrium**.

  After that, if the push force continues to increase, the object will begin to accelerate.
- F_{max} is proportional to R, so $F_{max} = \mu R$ where μ is the **coefficient of friction**.
- The rougher the surface, the higher the value of μ
- Frictional force can vary up to its maximum so $F \le \mu R$

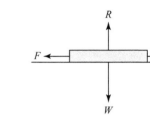

Example 1

A crate of mass 40 kg is being pushed up a rough ramp inclined at 30° to the horizontal by a force of 250 N.

Calculate the coefficient of friction if the crate is on the point of sliding up the ramp.

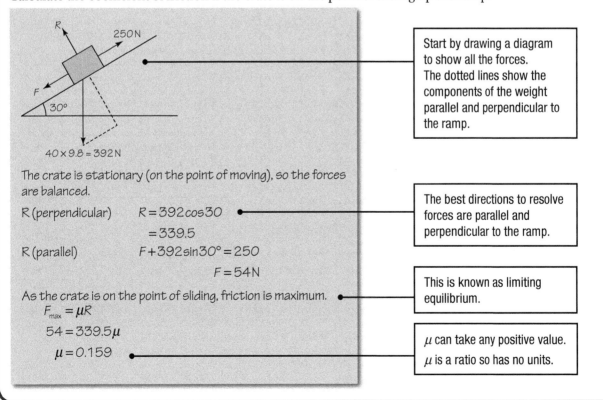

Start by drawing a diagram to show all the forces. The dotted lines show the components of the weight parallel and perpendicular to the ramp.

The crate is stationary (on the point of moving), so the forces are balanced.

R (perpendicular) $R = 392\cos 30$

$= 339.5$

R (parallel) $F + 392\sin 30° = 250$

$F = 54\,N$

The best directions to resolve forces are parallel and perpendicular to the ramp.

As the crate is on the point of sliding, friction is maximum.

$F_{max} = \mu R$

$54 = 339.5\mu$

$\mu = 0.159$

This is known as limiting equilibrium.

μ can take any positive value.
μ is a ratio so has no units.

Example 2

A block of mass M rests on a rough horizontal table with $\mu = \dfrac{1}{2}$

Mass M is attached to a light inextensible string making an angle of $30°$ to the horizontal. The string passes over a light, smooth pulley and is joined to a block of mass m hanging freely vertically below the pulley.

The system is on the verge of moving. Show that $\dfrac{M}{m} = \sqrt{3} + \dfrac{1}{2}$

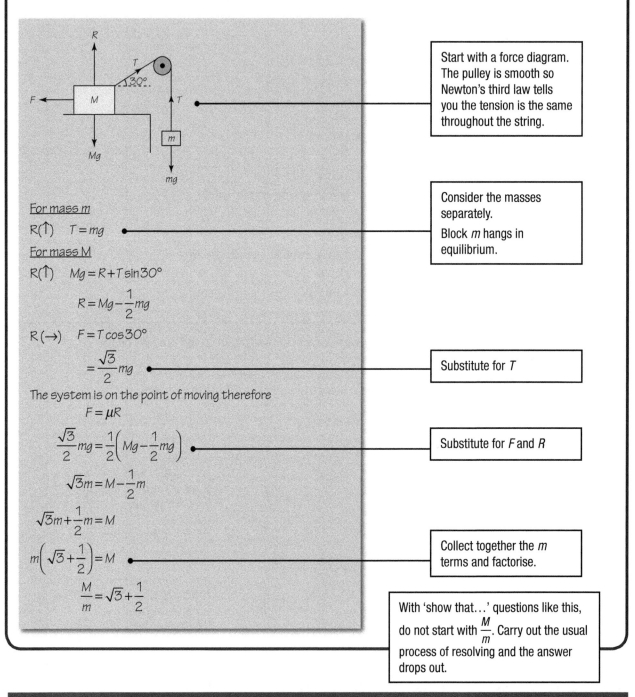

For mass m

$R(\uparrow) \quad T = mg$

For mass M

$R(\uparrow) \quad Mg = R + T\sin 30°$

$\qquad R = Mg - \dfrac{1}{2}mg$

$R(\rightarrow) \quad F = T\cos 30°$

$\qquad = \dfrac{\sqrt{3}}{2}mg$

The system is on the point of moving therefore

$\qquad F = \mu R$

$\dfrac{\sqrt{3}}{2}mg = \dfrac{1}{2}\left(Mg - \dfrac{1}{2}mg\right)$

$\sqrt{3}m = M - \dfrac{1}{2}m$

$\sqrt{3}m + \dfrac{1}{2}m = M$

$m\left(\sqrt{3} + \dfrac{1}{2}\right) = M$

$\dfrac{M}{m} = \sqrt{3} + \dfrac{1}{2}$

Start with a force diagram. The pulley is smooth so Newton's third law tells you the tension is the same throughout the string.

Consider the masses separately.
Block m hangs in equilibrium.

Substitute for T

Substitute for F and R

Collect together the m terms and factorise.

With 'show that…' questions like this, do not start with $\dfrac{M}{m}$. Carry out the usual process of resolving and the answer drops out.

Exam tips

- Start with a large, clearly drawn force diagram.
- Resolve in two directions at right angles to each other.
- Make it clear which body you are considering.
- Explain which forces you are considering and state their directions.

Exam practice questions

Where necessary take g as $9.8\,\mathrm{m\,s^{-2}}$

1 A box of mass $4\,\mathrm{kg}$ rests on rough ground with $\mu = 1.1$

    a Julie tries to move the box by pushing down on it with a force directed $25°$ below the horizontal.

       What is the smallest force she needs to get the box moving? [5]

1 b Marc pulls the box with a string angled 25° above the horizontal.

What is the smallest force he needs to get the box moving? [4]

2 A box of mass 6 kg rests on a rough table with $\mu = 0.6$. It is attached to masses of 5 kg and m kg by light inextensible strings that pass over smooth light pulleys as shown in the diagram.

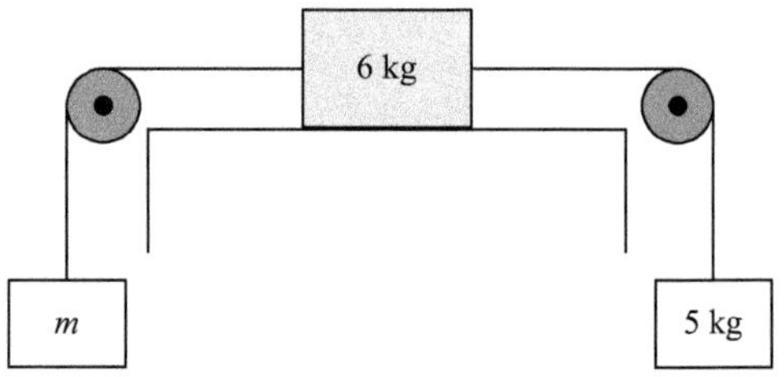

Given that the system remains at rest find the range of possible values of m. [6]

3 A mass of 5 kg rests on a rough horizontal table with coefficient of friction 0.3. It is connected by light inextensible strings that pass over smooth light pulleys to a mass of 8 kg that hangs freely and a mass of 10 kg that rests on a rough slope inclined at 30° to the horizontal. The coefficient of friction between the 10 kg mass and the slope is μ

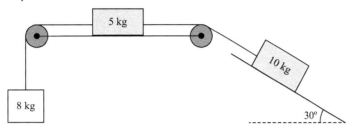

Given that the 10 kg mass is on the point of moving up the slope, determine the value of μ
Give your answer to 3 significant figures. [5]

Recap

- Dynamics is the study of acceleration and motion as a result of forces.
- You need to be confident about resolving vectors into components.
  It's easy to make mistakes here so a large clear diagram will help you.
- When solving problems the key skills are to
  - recognise any direction where there is *no acceleration* and balance forces
  - recognise any direction where there *is acceleration* and use Newton's second law.
    If one of the forces depends on velocity, v, it means the acceleration is variable.

    You should use $Force = m\dfrac{dv}{dt}$ for Newton's 2nd law.

    This could lead to a differential equation where you may need to separate the variables to solve.
    (Don't forget the constant of integration!)

Example 1

A box of mass 20 kg moves down a surface inclined at 30° to the horizontal.

The coefficient of friction is 0.2

Find the acceleration of the box. Use $g = 9.8\,\text{m s}^{-2}$

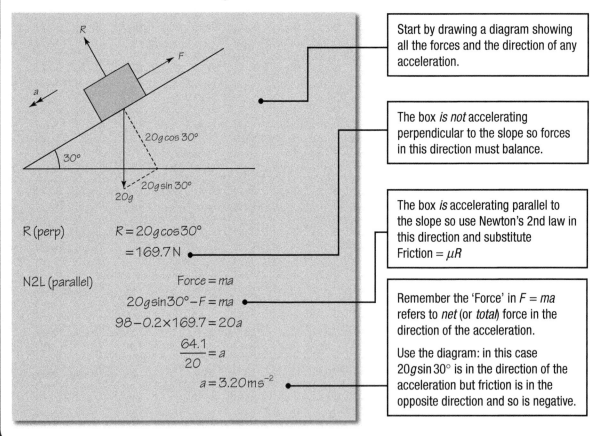

Start by drawing a diagram showing all the forces and the direction of any acceleration.

The box *is not* accelerating perpendicular to the slope so forces in this direction must balance.

$R\,(\text{perp})$ $R = 20g\cos 30°$
                $= 169.7\,\text{N}$

The box *is* accelerating parallel to the slope so use Newton's 2nd law in this direction and substitute Friction $= \mu R$

$N2L\,(\text{parallel})$ $Force = ma$
                 $20g\sin 30° - F = ma$
              $98 - 0.2 \times 169.7 = 20a$
                    $\dfrac{64.1}{20} = a$
                     $a = 3.20\,\text{ms}^{-2}$

Remember the 'Force' in $F = ma$ refers to *net* (or *total*) force in the direction of the acceleration.

Use the diagram: in this case $20g\sin 30°$ is in the direction of the acceleration but friction is in the opposite direction and so is negative.

Example 2

A car of mass 900 kg is towing a trailer of mass 600 kg. The driving force of the car is 4 kN. The total resistance on the car is 400 N and on the trailer it is 600 N

a Find the acceleration of the car and trailer.

b Find the force in the tow bar.
State whether it is a thrust or a tension.

> For connected bodies you can either consider the whole system or the individual bodies.

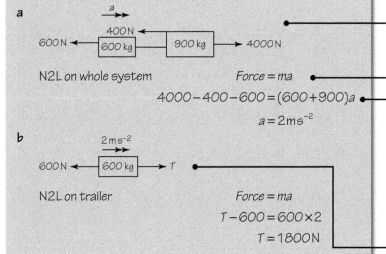

a

600 N ← 400 N ← [600 kg] [900 kg] → 4000 N

N2L on whole system   Force = ma

$$4000 - 400 - 600 = (600 + 900)a$$

$$a = 2\,\text{ms}^{-2}$$

> *Force* is the net force acting on the whole system i.e. the driving force – the resistances.

> Any force in the tow bar is an internal force so is we do not consider it when dealing with the whole system.

b

600 N ← [600 kg] → T   (2 ms⁻²)

N2L on trailer   Force = ma

$$T - 600 = 600 \times 2$$

$$T = 1800\,\text{N}$$

> Assume the force in the tow bar is a tension (pull).

> The positive answer tell you the assumption was correct. If it was wrong you would get a negative answer.

Example 3

A jet aircraft, of mass 5000 kg, flies horizontally at $v(t)$ m s⁻¹ after t seconds.
It produces 900 kN of thrust and experiences air resistance of $10v^2$ N

Show that $t = \dfrac{5}{6}\ln\dfrac{c(300+v)}{300-v}$

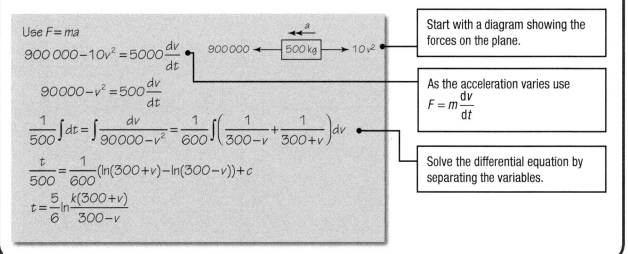

Use $F = ma$

$$900\,000 - 10v^2 = 5000\frac{dv}{dt}$$

$$90000 - v^2 = 500\frac{dv}{dt}$$

$$\frac{1}{500}\int dt = \int \frac{dv}{90000 - v^2} = \frac{1}{600}\int\left(\frac{1}{300-v} + \frac{1}{300+v}\right)dv$$

$$\frac{t}{500} = \frac{1}{600}(\ln(300+v) - \ln(300-v)) + c$$

$$t = \frac{5}{6}\ln\frac{k(300+v)}{300-v}$$

> Start with a diagram showing the forces on the plane.

> As the acceleration varies use $F = m\dfrac{dv}{dt}$

> Solve the differential equation by separating the variables.

Exam tips

- Draw a large, clear diagram.
- Mark all the forces on the diagram and any acceleration.
- Do not invent forces. For example, in Example 1 some students wrongly mark a force going down the slope as the box moves in that direction.
- Remember, in $F = ma$, where the force is the *net force* in the direction of acceleration.

Where necessary take g as $9.8\,\mathrm{m\,s^{-2}}$

1 A mass of 5 kg starts from rest on a slope inclined at 20° to the horizontal.
The coefficient of friction, $\mu = 0.1$

Find the acceleration of the mass and the time taken to get to a speed of $10\,\mathrm{m\,s^{-1}}$ [5]

2 A mass of M kg is attached to a mass of m kg, where $M > m$ by a light inextensible string that passes over a light smooth pulley.

Show that the acceleration of the system is $a = \dfrac{g(M-m)}{M+m}$ [4]

3 A mass of 5 kg rests on a rough surface inclined at 30° to the horizontal. It is attached to a mass of 1 kg by a light inextensible string that passes over a light smooth pulley at the top of the slope. The 1 kg mass hangs freely and the coefficient of friction between the 5 kg mass and the surface is $\mu = 0.2$

Calculate the acceleration of the system, stating whether the 5 kg block accelerates up or down the slope. [6]

4 A rocket-powered car of mass 7.8 tonnes is trying to beat the land speed record. Its engines can develop 212 kN of thrust and at velocity $v\,\text{m s}^{-1}$ it experiences a drag of $v^2\,\text{N}$

a Show that after the engines are turned off $\dfrac{1}{v} = \dfrac{t}{7800} + c$ [5]

4 **b** If, at the car's top speed of $470 \, \text{m s}^{-1}$, the engines are turned completely off, find how long it takes the car to decelerate to $355 \, \text{m s}^{-1}$ [3]

_____

_____

_____

_____

_____

_____

_____

5 Masses of 5 kg and 3 kg are attached by a light inextensible string that passes over a light smooth pulley.

The 5 kg mass rests on a rough surface with $\mu = 0.3$ inclined at $30°$ to the horizontal.

The 3 kg mass rests on a smooth slope inclined at $60°$ to the horizontal.

Show that the system is in equilibrium. [6]

_____

_____

_____

_____

_____

_____

_____

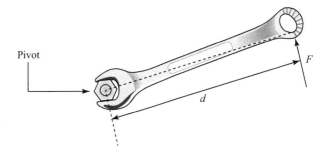

- A moment is the **turning effect** of a force about a **pivot point**.
  For example, pushing a spanner to turn a nut.
- This idea highlights two important ideas about moments.
    - The turning effect is proportional to the force (F) and to its perpendicular distance from the pivot point (d) leading to the formula

$$\boxed{\text{moment} = Fd}$$

    - The distance, d, is measured at right angles to the force. If you imagine trying to turn the nut you would push the spanner at right angles to get the maximum turning effect.
- The unit for moments is the Newton metre (N m)
- When the force is given in component form, consider each component separately and remember to find the perpendicular distance from the component to the pivot.
- Moments must be considered when working out whether an object is in equilibrium.
  For an object in equilibrium both the following conditions must be met.
    - The sum of the **forces** is 0
      All the forces in one direction equal all the forces in the opposite direction.
    - The sum of the **moments** is 0
      All the clockwise moments equal all of the anticlockwise moments.
- If the force is not at right angles to the distance from the pivot point there are two ways of thinking about the situation:
    - use your diagram to mark and find the perpendicular (shortest) distance from the line of action of the force to the pivot point
    - split the force into components parallel and perpendicular to the line of action of the force and use the perpendicular component
  Both calculations give the same answer but there can be contexts where one approach is much easier than the other.

The two ways to deal with this force F acting at angle θ are shown below.
Both methods give moment $= F\sin\theta \times d$

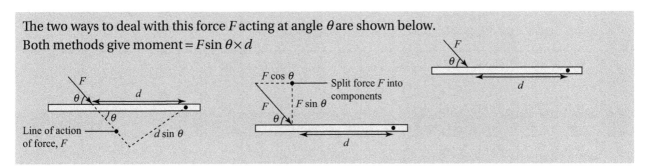

Example 1

A 2 m uniform rod, AB has mass 6 kg and forces of 30 N, 40 N and 50 N applied as shown in the diagram.

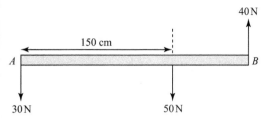

Find the total clockwise moment about point A

Force due to weight of rod = $6g$ N = $6 \times 9.8 = 58.8$ N

The rod is uniform so the weight acts through the midpoint.

A ⟳ $58.8 \times 1 + 50\sin 30 \times 1.5 - 40 \times 2$
$= 58.8 + 37.5 - 80$
$= 16.3$ Nm

The clockwise arrow shows that we are taking the clockwise direction as positive.

The 30 N force has no moment about A since $d = 0$

The 40 N force causes an anticlockwise motion and hence is a negative clockwise moment

Example 2

A rectangular lamina $ABCD$ with vertices $(0,0)$, $(0,5)$, $(8,0)$ and $(8,5)$ is acted on by the force $(4\mathbf{i} + \mathbf{j})$ N from the point with position vector $\mathbf{i} + 3\mathbf{j}$

A lamina is a 2D (flat) shape.

One unit represents 1 metre. Find the moment of this force about the point B

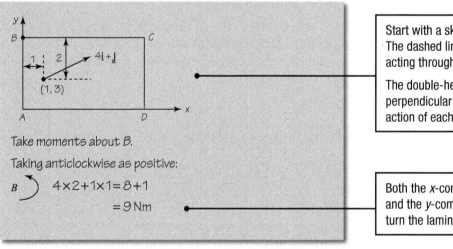

Take moments about B.

Taking anticlockwise as positive:

B ⟳ $4 \times 2 + 1 \times 1 = 8 + 1$
$= 9$ Nm

Start with a sketch.
The dashed lines show the components acting through $(1,3)$

The double-headed arrows show the perpendicular distances from the line of action of each component to B

Both the x-component of the force (4) and the y-component (1) are acting to turn the lamina anticlockwise about B

Exam tips

- Draw a large, clear diagram showing forces and distances.
- Remember that a moment is force $\times$ *perpendicular* distance.
- Think carefully about which point to take moments about so that the algebra is as easy as possible.
- Remember that clockwise and anticlockwise moments will have different signs.
- Take care not to mix up weight, which is a force, and mass which is a measure of a body's inertia.

1   What is the total clockwise moment about point *A*?

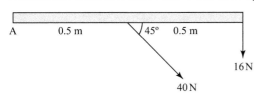

[2]

2   A uniform rod *AB* of length 2 m and weight 10 N is attached to a vertical wall at point *A*, where it can pivot freely, and supported in a horizontal position by a light, inextensible string attached to *B* and angled at 60° to the rod.

Take moments about A to find the tension in the string.   [4]

3 A force of $(2\mathbf{i}+5\mathbf{j})$N is applied at the point with position vector $\mathbf{i}+4\mathbf{j}$

Find the moment of this force about the point $4\mathbf{i}-2\mathbf{j}$ given 1 unit = 1 metre. [4]

----

4 A uniform rod AB of length 2 m and weight 20 N is supported in a horizontal position with supports at A and 1.5m from A

A weight of 4 N is attached to the rod at point B

Show that the forces from the supports are in the ratio $7:2$ [5]

----

5 A uniform circular table of radius a and weight W is supported by three vertical legs. The points A, B and C, where the legs meet the table, are the vertices of an equilateral triangle of side $\dfrac{\sqrt{3}}{2}a$

The centroid of this triangle is at point O, the centre of the table.

A weight P is placed at a point D on the edge of the table.

The plan view of this situation is shown in the diagram.

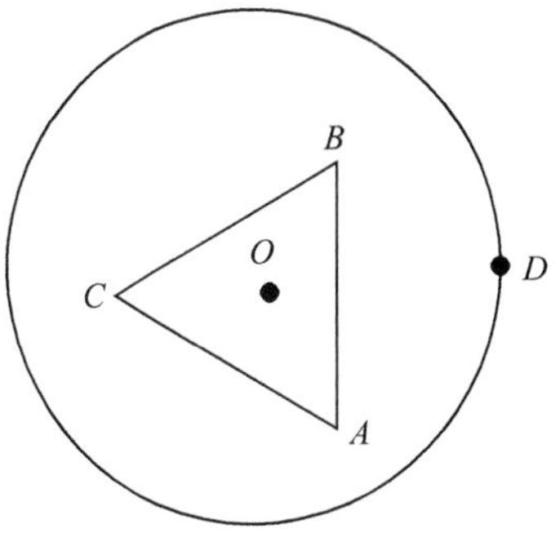

Find, in terms of W, the lowest value of P that will cause the table to tilt.

(Note that the centroid is $\dfrac{2}{3}$ of the length CE, where E is the midpoint of AB) [5]

Track your progress

Use these checklists to track your confidence level in each topic of A Level Mechanics

Ch	Objective	MyMaths	InvisiPen	Not yet	Almost	Yes!
7 Units and kinematics	Understand and use standard SI units and convert between them and other metric units.	2183	–	☐	☐	☐
	Calculate average speed and average velocity.	2183	07.1B	☐	☐	☐
	Draw and interpret graphs of displacement and velocity against time.	2183	07.2A	☐	☐	☐
	Derive and use the formulae for motion in a straight line with constant acceleration.	2184	07.3B	☐	☐	☐
	Use calculus to solve problems involving variable acceleration.	2289	07.4A	☐	☐	☐
8 Forces and Newton's laws	Resolve in two perpendicular directions for a particle in equilibrium.	2186, 2293	08.1A	☐	☐	☐
	Calculate the magnitude and direction of the resultant force acting on a particle.	2293	–	☐	☐	☐
	Resolve for a particle moving with constant acceleration. Work out acceleration of forces.	2187, 2293	08.2B	☐	☐	☐
	Understand the connection between the mass and the weight of an object. Know that weight changes depending on where the object is.	2185, 2187	08.3A	☐	☐	☐
	Resolve for "connected objects", such as an object in a lift.	2188	–	☐	☐	☐
	Resolve for particles moving with constant acceleration connected by string over pulleys.	2188	08.4B	☐	☐	☐
18 Motion in two dimensions	Use the constant acceleration equations for motion in two dimensions.	2290	18.1B	☐	☐	☐
	Use calculus to solve problems in two-dimensional motion with variable acceleration.	2291	18.2A	☐	☐	☐
	Solve problems involving the motion of a projectile under gravity.	2198, 2199	18.3B	☐	☐	☐
	Analyse the motion of an object in two dimensions under the action of a system of forces.	2192	18.4A	☐	☐	☐
19 Forces 2	Understand that there is a maximum value that the frictional force can take (μR) and that it takes this value when the object is moving or on the point of moving.	2193	–	☐	☐	☐
	Resolve in suitable directions to find unknown forces when the system is at rest or has constant acceleration.	2190, 2191	19.2A	☐	☐	☐
	Use constant acceleration formulae for problems involving blocks on slopes or blocks connected by pulleys.	2191	19.1B	☐	☐	☐
	Solve differential equations which arise from problems involving $F = ma$	2194	–	☐	☐	☐
	Take moments about suitable points and resolve in suitable directions to find unknown forces.	2197	19.3B	☐	☐	☐

Introduction to mechanics

Mathematical formulae for mechanics

Formulae that are provided

Kinematics

For motion in a straight line with constant acceleration:

$$v = u + at$$

$$s = ut + \frac{1}{2}at^2$$

$$s = vt - \frac{1}{2}at^2$$

$$v^2 = u^2 + 2as$$

$$s = \frac{1}{2}(u+v)t$$

Formulae that should be known

Mechanics

Forces and equilibrium

Weight $=$ mass $\times g$

Friction: $F \leq \mu R$

Newton's second law in the form: $F = ma$

Kinematics

For motion in a straight line with variable acceleration:

$$v = \frac{\mathrm{d}r}{\mathrm{d}t} \qquad a = \frac{\mathrm{d}v}{\mathrm{d}t} = \frac{\mathrm{d}^2 r}{\mathrm{d}t^2}$$

$$r = \int v \, \mathrm{d}t \qquad v = \int a \, \mathrm{d}t$$

Section 7.1

1 N, weight is a force measured in newtons.
2 $147 \, \text{m s}^{-1}$
3 $0.0612 \, \text{km h}^{-1}$
4 $1.02 \, \text{m s}^{-1}$
5 **a** $0.0694 \, \text{m s}^{-2}$ **b** $3.33 \, \text{km}$

Section 7.2

1 $-0.5 \, \text{m s}^{-1}$
2 **a**
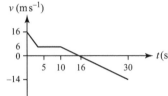
  b $5 \, \text{m}$
  c $0.167 \, \text{m s}^{-1}$
3 **a**
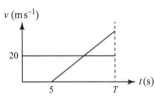
  b Car travels $20T$
    Motorcycle velocity at T is $2(T-5)$
    Motorcycle distance $\frac{1}{2}(T-5)2(T-5)$
    $20T = (T-5)(T-5)$
    $20T = T^2 - 10T + 25$
    $T^2 - 30T + 25 = 0$
  c $583 \, \text{m} \; (T = 29.1421...\text{s})$

Section 7.3

1 70
2 **a** $192 \, \text{m s}^{-1}$ **b** $6120 \, \text{m}$
3 $20 \, \text{m s}^{-1}$
4 **a** $324 = 8u + 32a$ **b** $u = 12.5 \, \text{m s}^{-1} \; a = 7 \, \text{m s}^{-2}$
5 $\text{gradient} = \dfrac{\text{vertical change}}{\text{horizontal change}}$
  $a = \dfrac{v-u}{t}$
  $at = v - u$
  $v = u + at$
6 $18 \, \text{m s}^{-1}$

Section 7.4

1 $20t^3 - 6t$
2 **a** $5 + 24t - 3t^2$ **b** $4 \, \text{s}$
3 $6 \, \text{s}$
4 **a** 121.6 **b** v is undefined when $t = 0$
5 $a = 3t^{-\frac{1}{2}}$
  $3t^{-\frac{1}{2}} = 1$
  $t = 9$
  $s = 4t^{\frac{3}{2}} + c$
  $t = 9, s = 100$
  $c = -8$
  $s = 4t^{\frac{3}{2}} - 8$
  $4T^{\frac{3}{2}} - 8 = \dfrac{3}{\sqrt{T}}$
  $4T^2 - 8\sqrt{T} = 3$
  $T^2 = \dfrac{3 + 8\sqrt{T}}{4}$

6 $v = \displaystyle\int a \, \mathrm{d}t$
  $= at + c$
  $t = 0, v = u$
  $v = u + at$
  $s = \displaystyle\int (u + at) \, \mathrm{d}t$
  $= ut + \dfrac{1}{2}at^2$
  Assuming $s = 0$ when $t = 0$

Section 8.1

1 $191°$
2 **a**
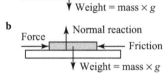
  b

3 **a** $50 \, \text{N}$, downwards   **b** $0 \, \text{N}$, in equilibrium   **c** $86.0 \, \text{N}$, at 54.5
4 14.6, at $15.9°$ above $\mathbf{i}$
5 $a = 5, b = 7$
6 $X = 9, Y = 4.5$

Section 8.2

1 $6 \, \text{m s}^{-2}$
2 **a** $3 \, \text{m s}^{-1}$ **b** $165 \, \text{s}$
3 $5.7°$
4 **a** $478 \, \text{N}$ **b** $722 \, \text{N}$

Section 8.3

1 $15 \, 500 \, \text{kg}, 55 \, 500 \, \text{N}$
2 $43.9 \, \text{kg}$
3 **a** $27 \, 440 \, \text{N}$ ($27 \, 500 \, \text{N}$ to 3 sf)
  b $12 \, \text{s}$
  c Light and inextensible
  d **i** $30 \, 240 \, \text{N}$ ($30 \, 300 \, \text{N}$ to 3 sf)
    ii $24 \, 640 \, \text{N}$ ($24 \, 700 \, \text{N}$ to 3 sf)
  e $13.9 \, \text{s}$
4. **a** $81 \, \text{kg}$ **b** $97.5 \, \text{kg}$ **c** $72.7 \, \text{kg}$ **d** $89.3 \, \text{kg}$

Section 8.4

1 $784 \, \text{N}$
2 **a** $1.225 \, \text{m s}^{-2}$
  b $T = 7a + 68.6$
    $T = 77.175$
    $= 77.2 \, \text{N}$
  c $33.1 \, \text{N}$
  d Pulley is frictionless and light; string is light and inextensible.
3 **a** $4000 \, \text{N}$ **b** $1500 \, \text{N}$, tension   **c** $2050 \, \text{N}$

Section 18.1

1 $(-0.25\mathbf{i} - \mathbf{j}) \, \text{m s}^{-2}$
2 $(30.9\mathbf{i} - 9.1\mathbf{j}) \, \text{m s}^{-1}$
3 $(2.2\mathbf{i} - 1.25\mathbf{j}) \, \text{m s}^{-2}$
4 9 seconds
5 $\mathbf{v} = (4\mathbf{i} + 6\mathbf{j}) + 2(4\mathbf{i} - 7\mathbf{j})$
  $= (12\mathbf{i} - 8\mathbf{j})$
  $|\mathbf{u}| = \sqrt{16 + 36}$
  $= \sqrt{52}$
  $|\mathbf{v}| = \sqrt{144 + 64}$
  $= \sqrt{208}$
  $= \sqrt{4} \times \sqrt{52}$
  $= 2\sqrt{52}$
  $= 2|\mathbf{u}|$
  i.e. final speed is double initial speed

6 $(-\mathbf{i} + 2\mathbf{j})\,\mathrm{m\,s^{-2}}$ or $(-2\mathbf{i} - \mathbf{j})\,\mathrm{m\,s^{-2}}$

Section 18.2

1 $(40\mathbf{i} - 2\mathbf{j})\,\mathrm{m\,s^{-2}}$

2 $\left(54\mathbf{i} - \dfrac{1}{108}\mathbf{j}\right)\mathrm{m\,s^{-2}}$

3 $(8\mathbf{i} - 9\mathbf{j})$

4 **a** $\left(\left(0.5e^{4t} - 0.5\right)\mathbf{i} + \left(0.2e^{2t} - 0.2\right)\mathbf{j}\right)\mathrm{m}$

    b $\dfrac{1}{2}e^{4t} - \dfrac{1}{2} = 0.2e^{2t} - 0.2$

    $5e^{4t} - 2e^{2t} - 3 = 0$

$$e^{2t} = \frac{2 \pm 8}{10}$$
$$= 1\ \mathrm{or} - 0.6$$

Implies $t = 0$ at origin so not NE or no solution. That is, no time when particle is NE

5 **a** $\mathbf{v} = 2\cos 2t\mathbf{i} - 2\sin 2t\mathbf{j}$

    $\mathbf{a} = -4\sin 2t\mathbf{i} - 4\cos 2t\mathbf{j}$

    $= -4(\sin 2t\mathbf{i} - \cos 2t\mathbf{j})$

    $= -4\mathbf{s}$

    b $2\,\mathrm{m\,s^{-1}}$ in the x-direction

6 $(t\mathbf{i} + (t\ln t - 1)\mathbf{j})\mathrm{m}$

Section 18.3

1 $4.9t^2 - 12t + 3 = 0$

$t = 0.283$ or 2.166

So over 3 m for 1.88 s

2 **a** $u\sin\alpha - gt = 0$

$$t = \frac{u\sin\alpha}{g}$$
$$s = ut\sin\alpha - \frac{1}{2}gt^2$$
$$= u\frac{u\sin\alpha}{g}\sin\alpha - \frac{1}{2}g\frac{u^2\sin^2\alpha}{g^2}$$
$$= \frac{u^2\sin^2\alpha}{g} - \frac{1}{2}\frac{u^2\sin^2\alpha}{g}$$
$$= \frac{u^2\sin^2\alpha}{2g}$$

    b $9.09\,\mathrm{m\,s^{-2}}$

3 $15°$ and $75°$

4 **a** $x = ut\cos\alpha$

so $t = \dfrac{x}{u\cos\alpha}$

$$y = ut\sin\alpha - \frac{1}{2}gt^2$$

Now substitute for t,

$$y = ut\sin\alpha - \frac{1}{2}gt^2$$
$$= u\frac{x}{u\cos\alpha}\sin\alpha - \frac{1}{2}g\left(\frac{x}{u\cos\alpha}\right)^2$$
$$= x\tan\alpha - \frac{1}{2}g\frac{x^2}{u^2\cos^2\alpha}$$
$$y = x\tan\alpha - \frac{g\sec^2\alpha}{2u^2}x^2$$

    b $y = \dfrac{\sqrt{3}}{3}x - \dfrac{4g}{3 \times 2u^2}x^2$

$$\frac{dy}{dx} = \frac{\sqrt{3}}{3} - \frac{4g}{3u^2}x$$
$$= \frac{\sqrt{3}}{3} - \frac{4g}{3u^2} \times \frac{u^2(5\sqrt{3} - 3)}{20g\,5}$$
$$= \frac{5\sqrt{3} - 5\sqrt{3} + 3}{15}$$
$$= \frac{1}{5}$$

Section 18.4

1 $3.61\,\mathrm{m\,s^{-2}}$, at angle $33.7°$ with **i**-direction

2 **a** $1.90\,\mathrm{m\,s^{-2}}$, $000.917°$ **b** $0.152\mathrm{N}$ west

3 **a** $507\,\mathrm{m}$,

    b Assumed that acceleration is constant. The drag would have increased as the velocity increased, therefore the acceleration would have decreased and the distance travelled would have been less.

4 $34.9\,\mathrm{N}$

5 $(4\mathbf{i} - \mathbf{j})\,\mathrm{N}$

Section 19.1

1 **a** $97.7\,\mathrm{N}$

    b $31.4\,\mathrm{N}$

2 $1.4 \leq m \leq 8.6$

3 0.173

Section 19.2

1 $2.43\,\mathrm{m\,s^{-2}}$, $4.11\,\mathrm{s}$

2 $Mg - T = Ma$

$T - mg = ma$

$Mg - mg = Ma + ma$

$g(M - m) = a(M + m)$

$$a = \frac{g(M - m)}{M + m}$$

3 $1.04\,\mathrm{m\,s^{-2}}$ down the slope

4 **a** $-v^2 = 7800\dfrac{dv}{dt}$

$$\frac{dt}{7800} = -\frac{dv}{v^2}$$
$$\frac{1}{v} = \frac{t}{7800} + c$$

    b $5.38\,\mathrm{s}$

5 $F = \mu R$

$= 0.3 \times 5g\cos 30$

$$= \frac{3\sqrt{3}}{4}g$$

Assuming 5 kg slides down

$$5g\sin 30 - T - \frac{3\sqrt{3}}{4}g = 5a$$
$$T - \frac{3\sqrt{3}}{2}g = 3a$$
$$a = -1.40$$

Assuming 3 kg slides down

$$T - \frac{3\sqrt{3}}{4}g - \frac{5}{2}g = 5a$$
$$\frac{3\sqrt{3}}{2}g - T = 3a$$
$$a = -1.2$$

As it can't be accelerating in either direction it must be in equilibrium

Section 19.3

1 $30.1\,\mathrm{N\,m}$

2 $5.77\,\mathrm{N}$

3 $-27\,\mathrm{N\,m}$

4 $20 \times 1 + 4 \times 2 = Q \times 1.5$

$$28 = \frac{3}{2}Q$$
$$Q = \frac{56}{3}$$
$$P + \frac{56}{3} = 24$$
$$P = \frac{16}{3}$$
$$\frac{56}{3} : \frac{16}{3} = 56 : 16$$
$$= 7 : 2$$

5 $\dfrac{W}{3}$